Sustaining the Forests of the Pacific Coast

Edited by Debra J. Salazar and Donald K. Alper

Sustaining the Forests of the Pacific Coast: Forging Truces in the War in the Woods

UBCPress · Vancouver · Toronto

Printed in Canada on acid-free paper

ISBN 0-7748-0815-2 (hardcover)
ISBN 0-7748-0816-0 (paperback)

Canadian Cataloguing in Publication Data

Main entry under title:

Sustaining the forests of the Pacific Coast

Includes bibliographical references and index.
ISBN 0-7748-0815-2 (bound)
ISBN 0-7748-0816-0 (pbk.)

1. Forest management—Environmental Aspects—Northwest, Pacific.
2. Forest policy—British Columbia. 3. Forest policy—Northwest, Pacific.
I. Salazar, Debra J. II. Alper, Donald K.
SD144.A13S97 2000   333.75'09795   C00-910780-0

UBC Press acknowledges the financial support of the Government of Canada through the Book Publishing Industry Development Program (BPIDP) for our publishing activities.

Canadä

We also gratefully acknowledge the support of the Canada Council for the Arts for our publishing program, as well as the support of the British Columbia Arts Council.

Set in Stone Serif by Bamboo and Silk Design Inc.
Printed and bound in Canada by Friesens
Copy editor: Dallas Harrison
Proofreader: Jacqueline Wood
Indexer: Patricia Buchanan

UBC Press
University of British Columbia
2029 West Mall, Vancouver, BC V6T 1Z2
(604) 822-5959
Fax: (604) 822-6083
E-mail: info@ubcpress.ubc.ca
www.ubcpress.ubc.ca

Contents

Preface / vii

Part 1: Introduction / 1

1 Politics, Policy, and the War in the Woods / 3
Debra J. Salazar and Donald K. Alper

Part 2: Institutions / 23

2 How the Way We Make Policy Governs the Policy We Make / 26
George Hoberg

3 International Dynamics of North American Forest Policy:
From Bilateral toGlobal Perspectives / 54
Thomas R. Waggener

4 Firms' Responses to External Pressures for Sustainable Forest
Management in British Columbia and the US Pacific Northwest / 80
Benjamin Cashore, Ilan Vertinsky, and Rachana Raizada

Part 3: Voices / 121

5 Forest People: First Nations Lead the Way toward a Sustainable
Future / 123
David R. Boyd and Terri-Lynn Williams-Davidson

6 The Multi-Ethnic, Nontimber Forest Workforce in the Pacific Northwest:
Reconceiving the Players in Forest Management / 148
Beverly A. Brown

Part 4: Policy Innovations / 171

7 A Crossroad in the Forest: The Path to a Sustainable Forest Sector in British
Columbia / 174
Clark S. Binkley

8 Wildlife Conservation on Private Lands: Habitat Planning and Regulatory Certainty / 193
R. Neal Wilkins

9 Multistakeholder Processes: Activist Containment versus Grassroots Mobilization / 209
Mae Burrows

Part 5: Conclusion / 229

10 Digging Out of the Trenches / 231
Debra J. Salazar and Donald K. Alper

Contributors / 244

Index / 247

Preface

In the spring of 1997 we organized a lecture series on forest politics and policy on the Pacific coast. We had planned the series for almost a year, attempting to provide a forum for voices that had largely been excluded from forest policy discourse. We were particularly interested in expanding the scope of forest policy conversation to include women, Native peoples, labour (including the most vulnerable workers who lack protection of the law), community activists, and smaller forestry enterprises. The lecture series was successful in stimulating conversation on our campus and in getting students to think beyond a unidimensional and polarized model of forest politics. So we committed ourselves to converting the lectures into a book in the hopes of providing a broader audience for the participants in the series.

The editors, both political scientists, come to this project from different backgrounds as well as shared perspectives and commitments. Debra Salazar was trained as a forest policy analyst, and though her current research and teaching interests extend well beyond forest issues, forests (and forestry) continue to occupy a special place for her. During the last several years, she has devoted substantial effort to integrating consideration of social justice into forest policy debate. To the extent that it focuses attention on those who lack political economic power, this book provides an opportunity to continue that effort. Don Alper's research interests focus on BC politics and the relations between British Columbia and the Pacific Northwest states. His work in Canadian studies is undergirded by a strong commitment to comparative studies and to the conviction that both countries can learn a great deal from each other.

A strong belief in democracy has shaped our approach to this book. Our intent is not to advocate a particular approach to forest land use or management. Rather, our purpose is to expand the scope of the debate and thus facilitate more democratic decision making. Forests are our common heritage, regardless of legal ownership or political jurisdiction. Thus, we wish to analyze the institutions and processes that shape forest management so that they can be held accountable to the will of the people. The challenges facing forest policy makers today are formidable and include intense value conflicts, economic globalization, and increasing skepticism of government. Sustaining the forests of the Pacific coast will require *political* analysis – analysis that examines inequality and conflict while highlighting opportunities to reduce the former and confront the latter.

This book represents our effort to contribute to such analysis. The contributing authors offer perspectives on political economic institutions, the cultural roles of forests, the status of forest workers, the content and distribution of property rights, and the meaning of citizenship. Taken as a whole, we attempted to produce a book that supports the exercise of responsible citizenship and the accountability of public institutions. To the extent that democracy is enhanced, the prospects for sustaining the forests are strengthened.

In the course of this project we benefited from the support of numerous people and institutions. Financial support for the lecture series and publication was provided by grants from the Canadian Embassy – Government of Canada/Gouvernement du Canada; the US Department of Education; and the Center for Canadian-American Studies, the Bureau for Faculty Research, the College of Arts and Sciences, the Environmental Center, Huxley College of Environmental Studies, and the Department of Political Science, all at Western Washington University. Geri Walker of the Bureau for Faculty Research was instrumental in facilitating our access to various funding sources. Our editors at UBC Press, Randy Schmidt and Ann Macklem, have been patient and supportive, while providing solid counsel.

Finally we have had the support of an awesome group of people at the Center for Canadian-American Studies at Western Washington University. Marty Hitchcock spent numerous hours (and days) helping to organize the lecture series, managing manuscript files, and coordinating just about every aspect of the project. Whether the task was chasing down a chapter author, digging up a reference, or copy editing a passage, our research assistants, Kristen Clapper, J.J. Hewitt, and Shelby Smith, never failed us. We are especially indebted to J.J. Hewitt, who played a particularly important role in helping us edit chapters. J.J.'s contributions and commitment to the project have gone far beyond what we had any right to expect of a student assistant. We owe the Canada House crew a great deal of gratitude for their contributions to this book and, even more than this, a special thanks for the opportunity to work with such bright, motivated, supremely competent, and fun people.

Part 1: Introduction

1
Politics, Policy, and the War in the Woods

Debra J. Salazar and Donald K. Alper

As the new century begins, a war is being waged over the forests of British Columbia and the US Pacific Northwest. The clash between combatants, motivated by radically different worldviews and opposing interests, has seemed to be unending and has occasionally erupted in violence. It has given impetus to a polarized and unproductive political debate that has failed to do justice to the richness and beauty of the forests of this region. Our purpose in this book is to blur the battle lines and to offer a less polarized view of forest politics in the region. In doing so, we hope to clarify the bases for the war in these woods and to suggest alternative ways of seeing, discussing, and managing forestry conflict.

While numerous authors[1] have examined the sources of forestry conflict, this collection offers two unique contributions. First, it examines forest policy in the context of a binational region, highlighting how two political systems facing similar cultural and economic challenges have treated their forests. During the past fifty years, British Columbia has constructed a system of expansive public ownership of forest land yet has delegated much managerial control to private entities (Wilson 1998). In contrast, the United States has developed a system of mixed ownership and a tradition of public involvement, even in decisions about private forest land (Dana and Fairfax 1980). The contributors to this collection analyze how these two systems have responded to demands associated with increased competition in global markets for timber and wood products, new findings in environmental research, and growing political pressure for protection of forest environments. Thus, readers are offered views of how political and economic institutions shape conflict and respond to pressure for change. The book examines the extent to which policy makers are constrained by institutional structures and global forces and how these institutions can be altered to serve new public ends.

Second, this book contributes to the forest politics discussion by highlighting the perspectives of those whose voices are often unheard in forest policy debate. Many observers characterize forest/environmental conflict as two-sided.[2] The inclusion of Indigenous people, nonindustrial owners of forest land, community-based activists, and nontimber forest workers makes the picture more complex. Adding these perspectives makes the lines of cleavage less sharp and offers the possibility of breaking open what too often has been a one-dimensional and polarized set of arguments.

It is conventional to see the war in the woods as pitting two coalitions against one another. On one side are forest industry officials and workers, rural communities, and forestry professionals. These sectors are seen as united in their defence of agro-industrial forestry, private property rights, and a liberal capitalist political economy. Across the barricades are environmentalists, both those who reside in urban areas and those who have moved back to the land. They promote ecoforestry, communal forms of land tenure, and participatory democratic institutions.

The combatants are thought to offer radically different visions of how our forests should look and which functions they should perform. Should they be vast green factories, producing fibre at ever faster rates? This agro-industrial vision relies on the use of technology to increase the efficiency of fibre production and to promote the region's competitiveness in global markets. This vision guided commercial forest management in North America for much of the last half of the twentieth century.[3] In part, Clark Binkley's analysis (Chapter 7) of the evolution of forest exploitation fits within this tradition. Binkley calls for zoning forests so that a small portion of the land base can be managed intensively to produce wood, while the remainder can provide other ecosystem services. His call to segregate production into land-use zones elaborates the agro-industrial model to accommodate the increasing and conflicting demands on the forest.[4] An alternative vision is of an ecological forestry that sees the forest as a home (Drengson and Taylor 1997; M'Gonigle and Parfitt 1994). According to this view, tending the forest should make it a safe home for all the species found there, from marbled murrelets to organic farmers. Biological diversity is the central value. Proponents of ecoforestry argue for vast reserves, or wild areas, that will not be subject to resource extraction.

This depiction of the conflict is accurate to some extent. There are two competing and polarized visions. But this portrayal glosses over important differences between actors in the two coalitions and leaves out other actors altogether. Even though observers often speak of the forest industry, or the timber sector, the nature of one's political economic claim on the forest shapes one's perspective on policy. In the United States, the policy interests of those who own forest land are not necessarily congruent with those of organizations that own manufacturing facilities. Neal Wilkins (Chapter 8) describes how a land-based company has responded to federal requirements to protect the habitats of endangered species. Such companies have little stake in the level of timber cut allocated to federal forests but are very much engaged in policy making related to private land. Similarly, independent loggers in British Columbia do not necessarily share policy interests with multinational paper companies. Lumping such disparate actors into a single category results in simplistic analysis and polarized politics.

Even within the relatively narrow sector of companies with global investments in land and processing facilities, important political and policy differences have emerged. Ben Cashore, Ilan Vertinsky, and Rachana Raizada (Chapter 4) take this observation further by examining three forest companies' responses to environmentalist pressure. These authors illustrate the importance of intrasectoral analysis; political strategies and public policies that treat forest companies uniformly will miss the leverage that can be obtained by recognizing their differences. Tom Waggener's chapter (3) highlights the global economic forces that are reshaping the political economy of forests on both sides of the forty-ninth parallel. Waggener notes that, although many of these changes will have similar consequences in both jurisdictions, others will have different effects.

Those left out of the one-dimensional view of forest politics include Indigenous peoples, who have articulated a model of forestry with both economic development and cultural integrity as goals (Scientific Panel for Sustainable Forest Practices in Clayoquot Sound 1995). While some Indigenous people have taken positions on each side of the forest divide, others have transcended it (Nathan 1993). But perhaps the key priority of Indigenous leaders, which is apparent in David Boyd and Terri-Lynn Williams-Davidson's chapter (5), is sovereignty. Although Boyd and Williams-Davidson argue that Native control of forests will promote a more sustainable and ecological forestry,[5] they do not condition their call for Native ownership rights on compliance with ecological goals.

Similarly, workers in what Beverly Brown (Chapter 6) terms the forest-floor economy have complex interests in the management of trees. Brown argues that efforts to achieve more local control of forest management[6] in the Pacific Northwest may further marginalize nontimber forest workers. Because many of these workers are recent immigrants, and many are not tied to geographically defined communities, local decision-making processes may exclude their participation and neglect their interests. If these workers' interests are to be addressed, then policy makers must find ways to facilitate their participation. Brown argues forcefully that failure to include nontimber forest workers will perpetuate injustice and undermine democracy.

Finally, many residents of timber-dependent rural communities are adopting new approaches in trying to forge truces in the war in the woods. There are many examples of multistakeholder deliberative processes in the region. Citizen-initiated processes include the Collaborative Learning Circle (southwest Oregon and northern California), the Quincy Library Group (northern California), and the Skagit Watershed Council (northwest Washington). These networks of community-based organizations demonstrate that environmentalists, fishers, and timber industry managers and workers can agree on forest practices in specific cases if not on the general principles that ought to guide forest management. The common thread in these processes is the belief that communities are threatened and that inclusive, community-based democratic processes should result in solutions. The conversation in which participants in these community-initiated processes have engaged illustrates that discourse about forests can be multidimensional and that actual forest practices can reconcile conflicting abstractions.

In contrast to citizen-initiated multistakeholder processes are those organized by governments. Oregon's watershed councils are a state response to concerns about degradation of salmon habitat. British Columbia's Commission on Resources and the Environment (CORE) was a provincial response to forest conflict. In Chapter 9, Mae Burrows examines the latter and argues that these processes, although viewed as open and inclusive, perpetuate inequality in the distribution of power.

It is clear that, if the forests of the Pacific coast are to be sustained, then the preferences of numerous actors will have to be integrated into policy making. For this to happen, inequality and conflict must be confronted. Romm (1993) has argued that sustainability is a political concept. What is to be sustained is determined by social values, and in nearly every management context there will be conflicting values. Some people will want forests to sustain outputs of commodity resources; others will focus on sustaining wildlife, recreational opportunities, or ecosystem processes. Moreover, the extent to which various actors are able to achieve, or even voice, their preferences depends on the distribution of political power. It is in confronting both conflict and inequality that our political institutions gain the capacity to become instruments of sustainability.

The contributors to this volume offer perspectives on forest conflict and sustainability that recognize the complexity of the issues and the diversity of interests in forest politics. These analyses suggest that hope may be found in recognizing cross-cutting values and interests as well as in being willing to imagine alternative, and often small-scale, democratic experiments. But conflict retains a central place in these analyses. Moreover, conflict has developed in two jurisdictions with different political systems but with similar cultural values and forest ecosystems. But conflict is not contained within state, provincial, or national jurisdictions, since many of the key actors operate in binational and international contexts and since global economic forces influence the terms of debate. Thus, although forest politics is increasingly played out across international boundaries, public policy solutions are constrained and channelled by domestic political systems. In the remainder of this chapter, we focus on the key attributes of those political systems, on several issues that have provoked controversy, and on the ways in which governments have responded to them.

Institutional Setting

Although the actors and the issues are similar on both sides of the forty-ninth parallel, the systems of government are very different. Each form of government uniquely shapes the political environment in which forest politics takes place.

The Canadian parliamentary system concentrates power in the political executive, which consists of the prime minister and cabinet in the federal government and the premiers and cabinets in the provincial governments. Unlike in the United States, where the separation of powers invites rivalry and competition between the executive and legislative branches, this conflict is virtually nonexistent in the Canadian federal government and in the provinces. The parliamentary system is based on the principle of responsible government, in which cabinets are formed and survive based on majority support in the legislature. Following an election, the leader of the political party with the most seats in the legislature assumes the position of prime minister (premier in the provinces) and appoints a number of elected members of his or her party to the cabinet. The prime minister of Canada, the provincial premiers, and their respective cabinets are also members of the legislative body, sitting in on and directing the activities of the assemblies. If cabinet ministers lose the confidence of a majority in the legislature, then they are expected to resign, and an election is called. In practice, cabinets rarely lose the support of the legislature because of the discipline imposed on the members by political parties. Strong party discipline provides the prime minister and cabinet with the tools to effectively lead the government and dominate policy making.

The structure of government is very different in the United States. The congressional system divides power over policy making between rival branches: the executive and the legislature. Each branch's institutional power is entrenched by the Constitution. Independence of and competition between the branches is encouraged by separate systems of election and different terms of office. Unlike in the parliamentary system, power is diffused across the branches rather than concentrated in the executive. Relatively weak party discipline makes it impossible for the president in the national government or governors in states to dominate policy making. Members of Congress and state legislatures secure positions of power by chairing committees that control the substance and flow of legislation. Although the popularly elected president and state governors have a decided advantage in directing the policy process, the system of separated institutions sharing power puts a premium on negotiating as a political style. George Hoberg, in his chapter (2), highlights the importance of the separation of powers model in the United States in shaping a policy process that has been more open than the BC system to environmentalists' demands for preservation of old-growth forests.

State governors typically share power with powerful legislators and other independently elected state officials. The legislature represents a rival power centre, and in Washington and Oregon states some heads of executive departments are elected and often belong to a different party than the governor. Thus, although the governor holds centre stage, he or she is at best only first among equals (Scott 1992). In contrast, the premier in British Columbia is the undisputed head of the executive branch by virtue of the power to appoint all members of the cabinet. The absence of powerful institutions to check premiers and prime ministers in the Canadian system results in executive-centred government. Regardless of which party is in power, the premier and the cabinet have an exceptionally high degree of control over the province.

Both Canada and the United States are federal systems, but they have significant differences that bear directly on forest politics. In terms of each federal system, provinces in general are more powerful than states. In Canada, provincial ownership of natural resources ensures that provinces are the key players in resource policy. Provincial control over forest resources makes for a close relationship between the BC government and the forest industry with regard to access to and management of the resource. Federal government involvement is primarily in the trade of forest products, a critical part of the BC economy. Thus, the provincial government inevitably must be reconciled with federal international trade interests. In the United States, a significant portion of forested land is owned by the federal government. Hence, it is directly responsible for the management of national forests in states. As a result, agencies, congressional committees, and the courts have become directly involved in forest policy in Washington, Oregon, and other states that contain federal public land.

In Washington and Oregon, forest land is divided among federal, state, and private holdings. These differing patterns of land ownership make for quite different policy and management approaches on these lands. The differences and relationships among jurisdictional agendas and policies are important in structuring forest politics in the United States.

Political Culture Context

The two jurisdictions may also be compared in terms of political culture. We use the term political culture to refer to the broad pattern of values and attitudes that defines the conditions for political action and guides and gives meaning to political behaviour.[7] Although political culture consists of largely unspoken attitudes, they affect how citizens make demands on the system and determine their orientations toward the decision-making apparatus, policy outputs, and the overall political community (Dyck 1993).

Research has shown that populism and polarization are significant features of BC political culture (Blake, Guppy, and Urmetzer 1996-97). Populism refers to the belief that "the people" should participate in political and economic decisions that will affect their lives. Embedded in the populist mind-set is suspicion of bureaucrats and "experts," trust in popular government, and belief in grassroots participation in setting directions for public policy and morality (Elkins 1985: 62). The fact that populist sentiments are a significant part of the provincial pattern of values is advanced as a major explanation of the continuing success of nonestablishment political parties in British Columbia (Blake 1996). Populism is also a key element of environmentalism in the province (Salazar and Alper 1999). Environmentalists who portray government forest policy and the Ministry of Forests as unaccountable to the public draw on populist rhetoric.

Observers of BC politics have noted for years the severe ideological polarization in the province. This historical rivalry between right and left is rooted in the traditional resource-based economy and the class conflict that it spawned (Black 1968; Robin 1972). In recent decades, the rise of the service economy and a growing postmaterialist ethos centred on environmentalism have substantially transformed class divisions (Howlett and Brownsey 1992). However, the bipolar structure of the BC party system remains in place, and political discourse continues to reflect left/right rhetoric (Blake 1996). Thus, while environmentalists articulate a politics that looks past (some might say ignores) conventional class divisions, BC political parties continue to see a world where the key actors are business, labour, and the state (Salazar and Alper 1996).

In recent years, ecology-oriented values have become an important part of BC political culture (Blake 1996; Blake, Guppy, and Urmetzer 1996-97; Harrison 1996). Support for environmentalism transcends social classes but is disproportionately located among supporters of parties on the left. Environmentalism also invokes populist values such as a belief in direct citizen action and opposition to professional bureaucrats and politicians. Conflict over forest practices has strengthened environmental consciousness and pushed environmental issues to the forefront of the province's political agenda.

The political cultures of the states in the Pacific Northwest are infused with strong populist values as well as the ideas of the progressive movement, which set deep roots in the political systems of most western states in the early part of the twentieth century. Like populism, progressivism is deeply suspicious of political and economic elites. The view that "self-serving" governors and legislatures are dangerous and need to be buffered with popular decision-making processes is at the heart of progressivism. In both Washington and Oregon, the progressive influence is reflected in public utility districts, apolitical commissions and boards, and direct-democracy practices such as initiatives and referenda (Mason 1994; Treleaven 1997).

Progressive and populist beliefs differ on the role of experts. Progressivism maintains that politics can and should be transformed into "administration" in which decisions are made on the basis of expertise and not compromise and negotiation. Central to progressivist values is the belief that nonpartisan administration is preferable to partisan government. Progressivism's faith in expertise has shaped federal forest management in the United States (Hays 1959). The Forest Service, which manages the vast majority of federal forest land in the region, has a long tradition of autonomy from political control. But during the past several decades, environmentalists have challenged the values and practices of agency foresters (Twight and Lyden 1988). Much of the conflict between these citizen activists and agency officials has its roots in the conflicting priorities of populism and progressivism.

Daniel Elazar (1984), in differentiating political cultures among states, identified Washington and Oregon as moralistic/individualistic political cultures. According to this categorization, the political system is seen as an arena for solving social problems that is also highly vulnerable to the machinations of powerful economic and political interests. Politics is seen ideally as a matter of concern for all citizens, not just those committed to professional political careers. This kind of political culture embraces the notion that every citizen should participate in civic activity. As a result, states with moralistic political cultures are more willing to place controls on interest groups and state officials and are more likely to adopt campaign finance and public disclosure restrictions (Walker 1992). These kinds of requirements have tended to make the political influence of the forest products industry more transparent.

In both Washington and Oregon, political parties are viewed with ambivalence. Both states have competitive party systems, but voters pride themselves on "voting for the person or issue" and not the party. This is consistent with a moralistic political culture in which parties are viewed as vehicles to attain goals believed to be in the public interest (Elazar 1984: 239). Washington is one of only a few states that do not restrict participation in primary elections for selecting party candidates to party members. The blanket primary allows voters to shift among the parties as they vote in nomination contests, and this approach further weakens the bond between voters and parties.

Support for environmental protection is also an important element in the political cultures of Washington and Oregon. Steel, List, and Shindler (1994) and Steel et al. (1995) have found that citizens in Seattle and Spokane, as well as throughout Oregon, have strong biocentric (ecology-oriented) attitudes with regard to forest policy. Consistent with these results, Ellis and Thompson's (1997) study of an Oregon county showed a substantial majority of citizens to be ecologically minded and engaged. These individuals were concerned about the possibility of an environmental catastrophe, unwilling to rely only on science and technology to prevent such catastrophic events, and supportive of expending more public resources to protect and improve the environment. These results from regional samples are consistent with the numerous national studies in the United States that identify the emergence of a broadly based environmental consciousness (Hays 1987; Jones and Dunlap 1992; Kanji 1996; Kempton, Boster, and Hartley 1995).

Key Issues in the Development of Forest Policy Conflict

Changes in cultural values and beliefs about natural resources have given rise to increasing public pressure to preserve and protect forest ecosystems. Respect for nonhuman nature, appreciation of natural beauty and integrity, concern about environmental health risks, and a sense of responsibility to future generations are widely held in North America (Kanji 1996; Kempton, Boster, and Hartley 1995). When these values filter public perceptions of industrial forest management, a picture of destruction and disregard emerges. It is this picture that has motivated many to call for radical changes in forest policy and management.

While the specifics of policy proposals differ on each side of the border, the general outlines of change are similar. Below we describe some of the key issues in the development of forest policy conflict in British Columbia and then in the US Pacific Northwest.

Clear-Cuts in the BC Wilderness[8]

In British Columbia, 94 percent of forest land is provincially owned and is referred to as Crown lands (Council of Forest Industries 1991). The provincial government licenses private companies to harvest trees from Crown forests through a tenure system. This system has been in place since the 1940s, though some licensing arrangements have been modified (Marchak 1983; Wilson 1998).[9] Licences vary with respect to length, renewability, and the distribution of responsibility for various management functions. Licensees make decisions about how, when, and how much to harvest within broad constraints. Integrated wood products companies control licences to 85 percent of the annual allowable cut in Crown forests (Forest Resources Commission 1991).[10]

During the 1980s, several issues provoked broad public discussion of forest policy in British Columbia and set the agenda for almost two decades of political conflict. The first issue was a reforestation backlog on Crown lands during the early 1980s. The second was prompted by Native efforts to settle land claims. The third was the "valley-by-valley" confrontation over clear-cut logging of ancient, or old-growth, forests.

The reforestation backlog became public as the forest industry headed into a severe recession. Forest managers and lay citizens criticized the government for insufficient investment in replanting clear-cut forests. While the reforestation backlog did not generate the same kind of public involvement that subsequent issues did, the problem indicated that the credibility of the forest management system in British Columbia was increasingly in question.

Native land claims and environmentalists' efforts to preserve ancient forests were intertwined in many of the most contentious forest battles of the 1980s and 1990s. Natives and environmentalists shared an immediate interest in stopping company logging, but their long-term interests often differed. In the Queen Charlotte Islands, local Haida people and environmentalists collaborated to stop logging on the South Moresby archipelago. The Haida acted to defend a land claim; environmentalists wanted to create a wilderness preserve. Both groups participated in litigation followed by other tactics to draw national and international attention to the islands. The Haida blockaded Lyell Island to prevent logging. Environmental organizations lobbied and tried to focus public attention on the Haida. In 1987, the federal government bought the cutting rights, and the federal and provincial governments agreed to create the South Moresby National Park Reserve. The Haida protested the creation of the park because their land claim remained unresolved.

The South Moresby battle was similar to confrontations throughout the province during the 1980s and 1990s. Environmentalists and Indigenous people engaged in litigation, lobbying, and direct action to stop logging of ancient forests. In some areas, environmentalists chained themselves to machinery and camped on platforms placed in trees. Perhaps the most prominent mass action took place in Clayoquot Sound during the summer of 1993, when hundreds of citizens were arrested at blockades for attempting to stop logging. These actions drew international attention to British Columbia's forests and pushed the provincial government to create forest reserves and to modify management practices in other forests.

The New Democratic Party (NDP) government, which came to power in 1991, took several steps to stop the war in the woods. First, the government recognized Aboriginal title and the inherent right to self-government, thus committing the province to negotiate land claims. Second, in 1992, the premier created the Commission on Resources and the Environment (CORE) to conduct regional land-use planning. At the same time, the minister of forests announced efforts to develop a new Forest Practices Code to regulate harvesting. Third, in 1993, the government initiated a Forest Sector Strategy to plan the industrial development of the forest sector.

These changes were designed to address three issues at the centre of the forest policy debate in British Columbia: the manner in which forests should be harvested, which lands should be subject to timber harvesting, and who should make decisions about forest management.[11] The government attempted to resolve the first issue through the Forest Practices Code,[12] enacted in 1994 and revised in 1997. The code creates a structure for the development and enforcement of regional and provincial forest practice regulations (Ministry of Forests 1994). This regulatory structure has received considerable criticism (Cook 1998). Forest operators have criticized the code for its extensive procedural requirements and the costs of compliance. Environmentalists have criticized the discretionary nature of most provisions and the lack of public access to the regulatory process. Struggles over the regulation of harvesting practices are likely to continue.

The second issue was land-use allocation. In its campaign platform, the NDP promised to designate 12 percent of the province's forest lands as "protected ecosystems." The pressure to decide which wild lands to protect was strong because of strains within the NDP constituency, which includes both environmentally minded citizens and wood products labour unions. Once in power, the NDP created CORE to conduct regional land-use planning.

CORE represented a move away from the old approach of ministry edicts toward a new participatory model emphasizing open discussion and the accommodation of different values. CORE established a process in which the province was divided into geographic areas, and the major economic, social, cultural, and governmental sectors in each area sent representative "stakeholders" to the table. The goal was to create a decision-making process that was local, inclusive, and consensual. It was hoped that, by including all stakeholders and providing the resources to develop a plan, the end result would incorporate the full range of values and therefore have greater credibility with the public. Mae Burrows argues in Chapter 9 that CORE failed to be inclusive and that its conception and design ensured that powerful sectors would retain their power and that community-based groups lacking funds and organization would remain disadvantaged. However, George Hoberg portrays CORE in Chapter 2 as a step toward opening up decision-making processes in the BC government and institutionalizing environmental values in forest policy.

The decision by the BC government to discontinue CORE in 1996 signalled diminished interest in shared decision-making approaches to resource issues. The disbanding of CORE also left the province without an overarching monitor of land-use practices (Owen 1998).

Notably, CORE did not address the third issue in the forest policy triad: forest tenure or control. That is, how should property rights be distributed among citizens, licence holders, First Nations, and the provincial government? Tenure issues were to be addressed, if at all, through the Forest Sector Strategy, which focused on the changing economic context of the industry. A twenty-member committee, composed primarily of officials from government, labour, and industry, was charged with developing the strategy.[13] Yet, as Boyd and Williams-Davidson point out in Chapter 5, recent court decisions support Aboriginal challenges to licence holders involved in logging on lands claimed by First Nations as traditional territories.

In sum, the provincial government has treated forest practice regulation as a technical issue and delegated its resolution to foresters and scientists. Land-use allocation has been managed as a public issue, with public participation invited through stakeholder negotiations. In contrast, tenure reform has been treated as a business issue, with the major forestry economic interests – industry, labour, and government – serving as the primary shapers of policy. The following chapters suggest that none of these structures is likely to contain the fundamental conflicts at the core of forest politics.

Endangered Species and Forests in Oregon and Washington

Unlike in British Columbia, the ownership pattern of US Pacific Northwest forests is varied. Roughly half of the forest land in Oregon and Washington is owned by the federal government (Waddell, Oswald, and Powell 1989).[14] Slightly less than 40 percent is privately owned. In Washington, the state controls 10 percent of forest land, while in Oregon state ownership accounts for 3 percent of forest land. Native American tribes control approximately 6 percent of forest land in Washington.

This mixture of ownership has given rise to a pattern of regulation in which both federal and state governments play major roles. The key laws governing forest management on federal lands include the National Forest Management Act (NFMA), the National Environmental Policy Act (NEPA), the Wilderness Act, and the Endangered Species Act (ESA). The first two laws govern planning, land-use allocation, and forest practices on federal lands. The third defines congressional authority to designate wilderness areas in which most forms of resource exploitation are proscribed. The ESA authorizes two cabinet departments to designate species as threatened or endangered and requires federal agencies to devise recovery plans for listed species. A key element of forest conflict in the Pacific Northwest has been associated with plans for the recovery of the northern spotted owl and more recently several subspecies of salmon.

State and private lands are subject to some federal statutes, most notably the ESA, and to a number of state laws. The ESA prohibits any landowner from harming a threatened or endangered species. The regulations implementing the law include land-use practices that might harm listed species. The extent to which the law authorizes the federal government to regulate private land use has been the subject of considerable controversy, resolved by a 1995 Supreme Court decision (*Babbitt* v. *Sweet Home,* 115 S.Ct. 2407). The court ruled that the Department of Interior could reasonably require landowners to modify their land-use practices to protect the habitat of designated endangered species.[15]

Both Oregon and Washington regulate private forestry through their forest practice acts. These laws authorize state agencies (the Department of Forestry in Oregon and the Department of Natural Resources in Washington) to adopt rules to protect public resources associated with private lands. Rules require reforestation after timber harvests and address the protection of soil, water, fish, wildlife, and other resources.

The use and management of forests have been contentious public issues in the Pacific Northwest for more than thirty years. While early conflicts focused on the designation of wilderness areas and the reduction of clear-cut sizes on federal lands, more recently forestry politics has shifted to a contest over the meaning of ecosystem management and the extent to which private lands should serve public ends.

The Politics of Ecosystem Management

In April 1993, just over three months after assuming office, President Clinton went to Portland for a promised forest summit. A series of federal court decisions during the late 1980s and early 1990s had nearly halted logging in the Pacific Northwest. Clinton had vowed to resolve the war in the woods by bringing contending parties together for dialogue that would lead to analysis. The summit took place over the course of one day. The more substantive part of the president's strategy was the appointment of a team of scientists to develop a set of policy options that would satisfy legal requirements for protection of endangered species' habitat and provide wood for the region's mills.[16] The name of the Forest Ecosystem Management Assessment Team (FEMAT) reflected the emerging consensus that forestry on federal lands should be guided by ecosystem management. But the FEMAT report as well as other conferences and reports have revealed considerable diversity of opinion with respect to the meaning of ecosystem management (Soden, Lamb, and Tennert 1998).

At an abstract level, this disagreement reflects contending positions regarding the ethical status of nonhuman nature (Normand and Salazar 1998). For many activists, scientists, and policy analysts, the central priority of ecosystem management is the protection of ecological integrity. This priority derives from both an ecocentric ethic and ecosystem science. Conservation biologists in particular have defined ecosystem management such that preservation of biological diversity is emphasized (Grumbine 1994). From this perspective, ecosystem management implies a public land system focused primarily on preservation of habitat. In contrast, many public land managers see ecosystem management as a tool for conflict management (Normand and Salazar 1998). For managers, ecosystem management connotes a forest management philosophy committed to the sustainable provision of a range of goods and services. From this perspective, ecosystem management may be used to build broadly based public support for land management plans that provide something for everyone.

In practical terms, contention regarding the meaning of ecosystem management is often reduced to decisions about timber extraction, where, how, and how many trees should be harvested (Calhoun 1998). Indeed, a prominent feature of the 1990s debate about federal forest policy involved contending parties using the authority of ecosystem management to support their positions on timber harvesting. George Hoberg's chapter (2) suggests how political institutional structure has contributed to the outcome of this debate. His analysis focuses on the preservation of old-growth forests, but his approach can certainly be applied to management issues on all parts of the public forest land base.

The Publicness of Private Forests
Private forests have also been contentious subjects of forest politics in the Pacific Northwest (Cashore 1998; Salazar and Cubbage 1990). There are few who challenge the premise that owners of private land have public responsibilities, though the nature and extent of those responsibilities are subject to debate.[17] Regulations for fire control and slash disposal have been in effect for decades. Since the 1970s, Oregon and Washington have regulated activity on private forests to protect water quality and fish habitat. Although there is little private forest land in British Columbia, the nature of tenure arrangements on public land and the use of the Forest Practices Code since 1995 to regulate the practices of tenure holders make for some useful comparisons.

Forest practice regulation on both sides of the border has been driven by arguments about the effects of logging on endangered species and their habitats. In British Columbia, large clear-cuts have made for important qualitative differences in the policy debate. But a key issue in all three jurisdictions – British Columbia, Washington, and Oregon – has been the extent to which the relevant government relies on standardized, nondiscretionary rules to achieve public ends or to create more flexible regulatory processes. In Chapter 8, Neal Wilkins examines one such process, the Habitat Conservation Plan. He draws on his experience as a wildlife biologist for a private timber company to argue that flexibility is essential to securing long-term protection of species and ecosystems. His analysis provides a point of departure for assessing the BC Forest Practices Code, which environmentalists have argued is so flexible as to be useless.

While policy analysts and political activists have focused on the government's role in securing private sector compliance, few have looked closely at how private forest companies respond to environmental regulation. The Cashore, Vertinsky, and Raizada chapter (4) begins to fill this gap in our understanding. Cashore and his co-authors develop a typology of company responses and suggest political and policy strategies for promoting environmentally sound behaviour by private firms.

Plan of the Book
The chapters in this collection explore multiple dimensions of forestry conflict in British Columbia, Washington, and Oregon. Organized in five sections, the chapters address key issues in the politics of forestry in the region. Part 2 describes global economic and domestic environmental/ political changes that have reframed regional conceptions of sustainability, examines how existing institutions have responded to these changes, and identifies factors that will affect future institutional responses. Part 3 considers the kinds of political and policy concerns raised by the perspectives of Indigenous peoples and nontimber forest workers. Chapters in this section raise important issues of equity, justice, and democracy. The authors of Part 4 explore how policy strategies of land-use zoning, cooperative business-government relations, and multistakeholder processes have been employed to address forest environmental issues. In the final chapter, the editors focus on four issues that consistently emerge in Parts 2, 3, and 4: monitoring, property rights, citizenship, and globalization.

Thus, the following chapters address questions of policy and politics and issues of pragmatic concern and normative commitment. As a whole, the collection analyzes the bases for resource conflict, identifies the institutional structures that circumscribe conflict resolution, outlines some emerging issues that policy makers will need to address, and offers some strategies for sustaining the forests of the region.

Notes
1 Recent book-length treatments include Chase (1995); Dietrich (1992); Harris (1995); Parfitt (1998); Wilson (1998); and Yaffee (1994).
2 Authors who portray the conflict as a clash between two sets of societal actors/interests include Burda, Gale, and M'Gonigle (1998); many of the contributors to Calhoun (1998); Chase (1995); and Wilson (1987-88, 1998).
3 See, for example, Day, Hart, and Milstein (1998), a study of Weyerhaeuser.
4 Rayner (1998) has noted that, although the idea of zoning has been proposed by those on opposing sides, their conceptions of the purpose of zoning differ. Some see zoning as a way to make forest production more efficient; others see it as a means to protect biological diversity. These two priorities lead to very different zoning maps.

5 Their conception of sustainable forest use is consistent with the model of ecoforestry outlined above.
6 Local control has been a key priority for some advocates of ecoforestry. See, for example, Burda, Gale, and M'Gonigle (1998).
7 See Salazar and Alper (1999) and Ellis and Thompson (1997) for more extended discussions of political culture and its role in environmental conflict.
8 The following historical sketch is derived largely from a review of the *Vancouver Sun* during the 1980s and 1990s and from accounts provided during our interviews with more than fifty participants in forest/environmental politics and policy making in British Columbia. Interviews were conducted between 1992 and 1994 and between 1996 and 1998. See Salazar and Alper (1996, 1999) for descriptions of interview subjects and questions.
9 See Marchak (1983) for a historical sketch of the development of tenure arrangements.
10 See Schwindt and Heaps (1996) for a detailed discussion of the distribution of costs and revenues associated with forest management in British Columbia.
11 See Leman (1988) and Tollefson (1998) for a more general analysis of forest policy in Canada.
12 Forest practice regulation constrains the processes of growing and harvesting wood to protect environmental values such as water quality and wildlife habitat (Salazar 1989).
13 Only one member of the committee had ties to environmental organizations in the province.
14 A smaller percentage of commercially available forest is controlled by the federal government, but this amount is in flux as decisions about land use continue to be made.
15 See Chapter 8 for a discussion of the implementation of the ESA on private lands.
16 For differing perspectives on the summit and the process that followed, see Chase (1995); Durbin (1996); and Rayner (1996).
17 In the United States, property rights activists have devoted considerable effort to limiting public regulation of private land use and to securing compensation for opportunities forgone as a result of regulation.

References

Black, E.R. 1968. "British Columbia: The Politics of Exploitation." In R. Shearer (ed.), *Exploiting Our Economic Potential: Public Policy and the British Columbia Economy*. 23-41. Toronto: Holt, Rinehart and Winston.

Blake, Donald E. 1996. "Value Conflicts in Lotusland: British Columbia Political Culture." In R.K. Carty (ed.), *Politics, Policy, and Government in British Columbia*. 3-17. Vancouver: UBC Press.

Blake, Donald E., Neil Guppy, and Peter Urmetzer. 1996-97. "Being Green in BC: Public Attitudes towards Environmental Issues." *BC Studies* 112: 41-61.

Burda, Cheri, Fred Gale, and Michael M'Gonigle. 1998. "Eco-Forestry Versus the State(us) Quo: Or Why Innovative Forestry is Neither Contemplated nor Permitted within the State Structure of British Columbia." *BC Studies* 119: 45-72.

Calhoun, John M. (ed.) 1998. *Forest Policy: Ready for Renaissance*. University of Washington, Institute of Forest Resources, Contribution No. 78. Seattle: University of Washington Press.

Cashore, Ben. 1998. "Governing Forestry: Environmental Group Influence in British Columbia and the U.S. Pacific Northwest." PhD diss., Department of Political Science, University of Toronto.

Chase, Alston. 1995. *In a Dark Wood: The Fight over Forests and the Rising Tyranny of Ecology*. Boston: Houghton Mifflin.

Cook, Tracey. 1998. "Sustainable Practices? An Analysis of BC's Forest Practices Code." In Chris Tollefson (ed.), *The Wealth of Forests: Markets, Regulation, and Sustainable Forestry*. 204-31. Vancouver: UBC Press.

Council of Forest Industries of British Columbia. 1991. *A Vision of BC Forests in the 21st Century*. Vancouver.

Dana, Samuel T., and Sally K. Fairfax. 1980. *Forest and Range Policy*. 2nd ed. New York: McGraw-Hill.

Day, Robert, Stuart Hart, and Mark Milstein. 1998. "Weyerhaeuser Forestry: The Wall of Wood." A Case Study from *The Business of Sustainable Forestry: Strategies for an Industry in Transition*, a project of The Sustainable Forestry Working Group. John D. and Catherine T. MacArthur Foundation.

Dietrich, William. 1992. *The Final Forest*. New York: Simon and Schuster.

Drengson, Alan, and Duncan Taylor. 1997. *Ecoforestry: The Art and Science of Sustainable Forest Use*. Gabriola Island, BC: New Society Publishers.

Durbin, Kathie. 1996. *Tree Huggers: Victory, Defeat, and Renewal in the Northwest Ancient Forest Campaign*. Seattle: The Mountaineers.

Dyck, Rand. 1993. *Canadian Politics: Critical Approaches*. Scarborough, ON: Nelson Canada.

Elazar, Daniel J. 1984. *American Federalism: A New View from the States*. New York: Harper and Row.

Elkins, David J. 1985. "British Columbia as a State of Mind." In Donald E. Blake (ed.), *Two Political Worlds: Parties and Voting in British Columbia*. 49-73. Vancouver: UBC Press.

Ellis, Richard J., and Fred Thompson. 1997. "Culture and the Environment in the Northwest." *American Political Science Review* 91(4): 885-97.

Forest Resources Commission. 1991. *The Future of Our Forests*. Victoria: Government of British Columbia.

Grumbine, R. Edward. 1994. "What Is Ecosystem Management?" *Conservation Biology* 8: 27-38.

Harris, David. 1995. *The Last Stand: The War between Wall Street and Main Street over California's Ancient Redwoods*. New York: Random House.

Harrison, Kathryn. 1996. "Environmental Protection in British Columbia: Post-Material Values, Organized Interests, and Party Politics." In R.K. Carty (ed.), *Politics, Policy, and Government in British Columbia*. 290-309. Vancouver: UBC Press.

Hays, Samuel P. 1959. *Conservation and the Gospel of Efficiency*. Cambridge, MA: Harvard University Press.

—. 1987. *Beauty, Health, and Permanence: Environmental Politics in the United States, 1955-1985*. Cambridge, UK: Cambridge University Press.

Howlett, Michael, and Keith Brownsey. 1992. "British Columbia: Public Sector Politics in a Rentier Resource Economy." In Keith Brownsey and Michael Howlett (eds.), *The Provincial State: Politics in Canada's Provinces and Territories*. 265-96. Toronto: Copp Clark Pittman.

Jones, Robert Emmett, and Riley E. Dunlap. 1992. "The Social Bases of Environmental Concern: Have They Changed over Time?" *Rural Sociology* 57(1): 28-47.

Kanji, Mebs. 1996. "North American Environmentalism and Political Integration." *American Review of Canadian Studies* 26(2): 183-204.

Kempton, Willett, James S. Boster, and Jennifer A. Hartley. 1995. *Environmental Values in American Culture*. Cambridge, MA: MIT Press.

Leman, Christopher K. 1988. "A Forest of Institutions: Patterns of Choice on North American Timberlands." In Elliot J. Feldman and Michael A. Goldberg (eds.), *Land Rites and Wrongs*. 149-200. Cambridge, MA: Lincoln Institute of Land Policy.

M'Gonigle, Michael, and Ben Parfitt. 1994. *Forestopia: A Practical Guide to the New Forest Economy*. Madeira Park, BC: Harbour Publishing.

Marchak, Patricia. 1983. *Green Gold: The Forest Industry in British Columbia*. Vancouver: UBC Press.

Mason, Thomas L. 1994. *Governing Oregon: An Inside Look at Politics in One American State*. Dubuque: Kendall-Hunt Publishing.

Ministry of Forests. 1994. *British Columbia Forest Practices Code: A Summary of Draft Regulations and Proposed Standards*. Victoria: Ministry of Forests.

Nathan, Holly. 1993. "Aboriginal Forestry: The Role of the First Nations." In Ken Drushka, Bob Nixon, and Ray Travers (eds.), *Touch Wood: BC Forests at the Crossroads*. 137-70. Madeira Park, BC: Harbour Publishing.

Normand, Valerie J., and Debra J. Salazar. 1998. "Assessing the Meaning of Ecosystem Management in the North Cascades." In Dennis L. Soden, Berton L. Lamb, and John R. Tennert (eds.), *Ecosystems Management: A Social Science Perspective.* Dubuque: Kendall-Hunt Publishing.

Owen, Stephen. 1998. "Land Use Planning in the 1990s: CORE Lessons." *Environments* 25(2-3): 14-26.

Parfitt, Ben. 1998. *Forest Follies: Adventures and Misadventures in the Great Canadian Forest.* Madeira Park, BC: Harbour Publishing.

Rayner, Jeremy. 1996. "Implementing Sustainability in West Coast Forests: CORE and FEMAT as Experiments in Process." *Journal of Canadian Studies* 31(1): 82-101.

—. 1998. "Priority-Use Zoning: Sustainable Solution or Symbolic Politics?" In Chris Tollefson (ed.), *The Wealth of Forests: Markets, Regulation, and Sustainable Forestry.* 232-54. Vancouver: UBC Press.

Robin, Martin. 1972. *The Company Province: The Rush for Spoils, 1871-1933.* Toronto: McClelland and Stewart.

Romm, Jeff. 1993. "Sustainable Forestry: An Adaptive Social Process." In Gregory H. Aplet et al. (eds.), *Defining Sustainable Forestry.* 280-93. Washington, DC: Island Press.

Salazar, Debra J. 1989. "Regulatory Politics and Environment: State Regulation of Logging Practices." *Research in Law and Economics* 12: 95-117.

Salazar, Debra J., and Donald K. Alper. 1996. "Perceptions of Power and the Management of Environmental Conflict: Forest Politics in British Columbia." *Social Science Journal* 33(4): 381-99.

—. 1999. "Beyond the Politics of Left and Right: Beliefs and Values of Environmental Activists in British Columbia." *BC Studies* 121: 5-34.

Salazar, Debra J., and Frederick W. Cubbage. 1990. "Regulating Private Forestry in the West and South." *Journal of Forestry* 88(1): 14-19.

Schwindt, Richard, and Terry Heaps. 1996. *Chopping Up the Money Tree: Distributing the Wealth from British Columbia's Forests.* Vancouver: The David Suzuki Foundation.

Scientific Panel for Sustainable Forest Practices in Clayoquot Sound. 1995. *Report 3: First Nations' Perspectives Relating to Forest Practice Standards in Clayoquot Sound.* Victoria: Scientific Panel for Sustainable Forest Practices in Clayoquot Sound.

Scott, George W. 1992. "The Office of Governor and Statewide Officials." In David C. Nice, J.C. Pierce, and Charles H. Sheldon (eds.), *Government and Politics in the Evergreen State.* Pullman, WA: Washington State University Press.

Soden, Dennis L., Berton L. Lamb, and John R. Tennert (eds.). 1998. *Ecosystems Management: A Social Science Perspective.* Dubuque: Kendall-Hunt Publishing.

Steel, Brent S., Peter List, and Bruce Shindler. 1994. "Conflicting Values about Federal Forests: A Comparison of National and Oregon Publics." *Society and Natural Resources* 7(2): 137-53.

Steel, Brent S., Nicholas P. Lovrich, John C. Pierce, Mary Ann E. Steger, and John Tennert. 1995. "Public Forest Policy Preferences in Washington and British Columbia: A Tale of Four Cities." Paper presented at the biennial conference of the Association for Canadian Studies in the United States, Seattle, 15-18 November.

Tollefson, Chris (ed.). 1998. *The Wealth of Forests: Markets, Regulation, and Sustainable Forestry.* Vancouver: UBC Press.

Treleaven, Michael S.J. 1997. *Social Policy and Regionalism in the Pacific Northwest.* Seattle: Canadian Studies Center, Henry M. Jackson School of International Studies.

Twight, Ben W., and Fremont J. Lyden. 1988. "Multiple Use vs. Organizational Commitment." *Forest Science* 34: 474-86.

Waddell, Karen L., Daniel D. Oswald, and Douglas S. Powell. 1989. *Forest Statistics of the United States, 1987.* USDA Forest Service, Pacific Northwest Research Station, Resource Bulletin PNW-RB-168. Portland: USDA Forest Service, Pacific Northwest Research Station.

Walker, Elizabeth. 1992. "Interest Groups in Washington State." In David C. Nice, J.C. Pierce, and Charles H. Sheldon (eds.), *Government and Politics in the Evergreen State.* 41-64. Pullman, WA: Washington State University Press.

Wilson, Jeremy. 1987-88. "Forest Conservation in British Columbia, 1935-1985: Reflections on a Barren Political Debate." *BC Studies* 76: 3-32.
—. 1998. *Talk and Log: Wilderness Politics in British Columbia*. Vancouver: UBC Press.
Yaffee, Steven. 1994. *The Wisdom of the Spotted Owl*. Washington, DC: Island Press.

Part 2: Institutions

British Columbia, Oregon, and Washington enjoy a common forest that is among the most productive in North America. Physically, the forest transcends the border, gradually blending into different forest types, reflecting local environmental factors. This forest, however, is bisected by an international border, and management has been shaped on each side by political and economic institutions. The institutions most directly engaged in forest management are government and forest companies.

Part 2 describes these institutions and the pressures confronting them. The cross-border perspective developed in this section focuses on how two political systems have treated the Pacific coast forest. During the past two decades, these systems have encountered similar challenges, one cultural, the other economic. The quality of the biophysical environment and the manner in which we treat our forests have become increasingly salient for citizens throughout the world and perhaps especially in this region. Those with environmental values argue for preservation of ancient, or old-growth, forests as well as for reformed management of second-growth forests. Changes in the international economy also present challenges to Canadian and US forest policy makers. Increasing globalization of corporate ownership and action have reshaped the policy environment, limiting the ability of national governments to regulate corporate behaviour effectively. Both Canadian and US governments have faced these challenges, one calling for radical change in the management and use of forest lands, the other constraining the nature and extent of change. By comparing the two countries' responses, we can gain insight into how two different sets of political economic institutions, which span one forest, have responded to environmentalism and globalization.

George Hoberg's chapter focuses on government and explores how two different policy systems shape forest politics. The US system is characterized by adversarial legalism, executive discretion, formal procedures, and judicial rulings. In contrast, the BC system grants far more discretion to executives and limits opportunities for citizens to challenge forest policies. Hoberg analyzes the effects of these contrasting policy styles on forest policies, including forest practices regulations and protection of old-growth forests.

Both systems have experienced strong pressures for change in recent years. Because policy making in both the US Pacific Northwest and British Columbia is embedded in larger political systems, the magnitude of potential change in policy styles is limited. Hoberg argues that the United States has been more successful in preserving old-growth forests but that forest management in British Columbia has changed substantially in response to pressure from environmentalists.

Although the political institutions of the two countries differ, from a global forest products perspective they act as a single region. North America represents the world's largest homogeneous market for forest products, leading to relatively standardized production processes and product standards in Canada and the United States. Canada, with large resources and a modest domestic marketplace, has found large and open markets south of the border. In turn, the United States, with a large population, has historically imported substantial quantities of forest products from Canada, supplementing domestic supplies of forest products to the economic advantage of consumers. But the bilateral forest relationship between Canada and the United States is under relentless pressure from globalization of the forest economy and the expansion of trade through both multilateral trade agreements.

Tom Waggener's chapter examines the international political economic context of forest policy in the region and argues that decisions made here may be local but that their causes and consequences are increasingly global. Waggener argues that the effects of national policies will increasingly depend on the often unpredictable responses of other actors in the global marketplace. Thus, US policies to restrict log exports did not result in the desired effect of increased lumber exports. Instead, other suppliers of raw logs met international demand. Similarly, Waggener contends, efforts to protect forest ecosystems in the United States and Canada by restricting harvests may give impetus to more environmentally destructive timber harvesting in other parts of the world.

Ben Cashore, Ilan Vertinsky, and Rachana Raizada examine how forest companies on both sides of the border have responded to domestic and international pressure from environmentalists. Unlike most analyses that explore the impact of environmentalism on public policy, their chapter focuses on firm responses to these social pressures. This neglected institution is important because a firm's activities are the ultimate target of environmental pressures even when these pressures are focused on the state. The important question, then, is whether firms are changing their behaviour in the direction of greater responsiveness to external pressures for environmental protection. If so, then what accounts for such change? This is especially important since environmental groups are increasingly bypassing the state by directly targeting firms through negative boycott campaigns and positive voluntary certification schemes. Cashore and his co-authors identify three types of response from firms (resistance, acquiescence, and proactive adoption of environmental practices) and the factors associated with each. In the process, the authors define a set of lessons for public policy makers, environmentalists, and forest company managers.

The following chapters thus address the effects of two sets of pressures (environmentalism and globalization) on two sets of institutions (government and forest companies). The analyses highlight both similarities and differences in institutional responses and identify relationships and trends that should interest those who care about forests and forestry in this region.

2
How the Way We Make Policy Governs the Policy We Make

George Hoberg

On both sides of the forty-ninth parallel, policy makers along the West Coast face similar forest policy problems. They confront exceptionally long time horizons and considerable factual uncertainty. They must choose how to allocate land among timber harvesting, wilderness preservation, and other conflicting uses. In areas where timber harvesting is to be allowed, they must decide how to regulate harvesting to protect nontimber values. Where jobs in the woods are declining, policy makers confront the difficult task of deciding how to support timber-dependent communities. These political issues are complicated by a complex spatial distribution of interests, in which residents of rural communities across the region depend on extractive activities such as logging for their livelihood, whereas much of the support for environmental preservation is in urban areas. The one major exception to this pattern of similar problems is that in the United States the issue of Aboriginal title has largely been settled for some time, while in British Columbia that process has just begun.

Despite these largely similar problems, policy makers across the forty-ninth parallel address them within quite different political systems. The BC government has its roots in the mother country of Great Britain, from which it has adopted the Westminster parliamentary system. Governments in the United States have their roots in a revolution against the eighteenth-century form of that British system, creating a distinctive system of separation of powers and checks and balances. These deep macropolitical differences have profound implications for how policy is made in the two countries, as well as significant impacts on the content and consequences of forest policies.

This chapter analyzes the structures and dynamics of the forest policy processes in British Columbia and the US Pacific Northwest[1] and examines their impact on one important area of forest policy: the preservation of old-growth forests. After providing a brief overview of the analytical framework, the chapter describes and compares the policy processes in the two jurisdictions. The chapter examines the processes by which old-growth forest has been set aside and then compares the amounts of old growth preserved to date in the two jurisdictions. It is not my purpose to argue a form of institutional determinism. In fact, my approach to explaining policy outcomes is resolutely multicausal. Nonetheless, a strong case can be made that institutional differences have had important policy consequences.

My approach to the study of policy is a *regime framework*. The policy regime framework consists of regime components, background conditions, and policy outcomes produced through the interactions of these two elements. The three components of policy regimes are actors, institutions, and ideas. Actors are the individuals and organizations, both public and private, that play important roles in the formulation and implementation of public policies. Actors have *interests* that they attempt to pursue through the political process, *resources* that they bring to bear in their efforts to influence public policy, and *strategies* to employ those resources in the pursuit of those interests. Institutions are rules and procedures that allocate authority over policy and structure relations between various actors in the policy process. Ideas are both causal and normative beliefs about the substance and process of public policy. For a specific policy area and a particular period of time, these three components combine to form a distinctive policy regime. Policy regimes exist in the context of certain background conditions, including public opinion, economic realities, and macropolitics. The interaction of regime components, in the context of particular background conditions, produces distinctive policy outcomes.

There is considerable disagreement in the political science literature about how much institutions matter.[2] There is widespread agreement, across different methodological perspectives, that institutions can affect the resources and especially the strategies of political actors. Some of the strongest advocates for an institutional approach also argue that institutions influence how actors define their interests. Most attempts to do so stress that institutions influence what actors come to believe is possible in terms of policy changes (e.g., Steinmo and Watts 1995). I think that this effect is more appropriately considered a matter of strategy and that it is better to conceive of interests as prior to and independent of institutional variables.

In forest policy, there are three primary categories of actors: a coalition of pro-logging groups intent on creating jobs and profits and minimizing the costs of environmental controls, a coalition of environmental groups interested in wilderness protection and stringent environmental controls on logging, and government. The third category is harder to define, in part because of the diversity of government actors and in part because of the differences across jurisdictions. It is reasonable to assume that the primary interest of elected politicians, both legislative and executive, is re-election. They may also be motivated to achieve good policy or institutional power, but in order to pursue these goals they must remain in office. Bureaucrats are assumed to be guided by personal incentives of income, power, and prestige; policy incentives; and organizational incentives of autonomy (Wilson 1989) and budget security and growth (Arnold 1979; Niskanen 1971). Judges are perhaps the most mysterious – they also have policy goals, but their overriding interest is to maintain the legitimacy of their actions (Caldiera 1991) by following precedents and adhering to the rule of law.

Institutions can influence policy through several avenues. First, they structure the authority and relations between government actors (e.g., between executive, legislative, and judicial, or between federal and subnational governments.) Second, they influence the relations between societal interests and the state. By specifying how interest groups can participate in the policy process, institutional rules shape the resources that interests can bring to bear and the strategies that they can adopt in pursuing those interests. More specific hypotheses about the impact of institutions will be presented after the two policy processes are compared.

The US Policy Regime: Pluralist Legalism
Ownership and governance of forests in the United States are complex. More than two-thirds of forested lands nationwide are in private ownership. Of the remaining amount owned publicly, most is controlled by the US federal government as part of the system of national forests under the jurisdiction of the US Forest Service, housed within the US Department of Agriculture. While they make up less land and produce less timber than private forests, the national forests are the most prized because of their diversity of habitat, wilderness values, and recreational opportunities. In the US Pacific Northwest, federal land ownership is higher than the national average – about 56 percent in the two states. On private lands, forestry is regulated by state governments. States regulate certain aspects of forestry, such as impacts on rivers and streams and pesticide use, but leave decisions about pricing and cut levels to companies and markets. On federal lands, forestry is regulated by a complex array of federal laws, summarized below. Harvesting is performed by private companies, which gain access to government timber through a competitive bidding process (Cashore 1999).

The battle over old-growth forests in the US Pacific Northwest has centred on the federal lands managed by the Forest Service or the Bureau of Land Management in the Department of Interior, in large part because virtually all of the old growth on private lands has already been harvested. As we will see, the fact that the relevant lands are controlled by the federal government in the United States is crucial to understanding the outcome south of the border and the differences between political dynamics in the two jurisdictions.

Like other areas of environmental policy, the forest policy regime in the United States is best characterized as "pluralist legalism" (Hoberg 1992, 1997a). There are three key features of this type of regime: (1) formal administrative procedures, with widespread access to information and rights to participation for all affected interests; (2) organized environmental groups with access to courts; and (3) nondiscretionary government duties, enforceable in court. This third characteristic is crucial because it grants private litigants a cause of action in court and gives the judiciary a legitimate role in scrutinizing agency decisions. Since the mid-1970s, US forest policy has displayed each of these features. This policy "style" emanates from the larger US political system dominated by separation of powers and checks and balances. Congress and the president have independent power and electoral bases and, for most of the time in the modern (post-1968) era, have been dominated by different parties. This division of powers fosters distrust between the branches, creating incentives for Congress to write statutes that place tight constraints on administrative discretion (Moe 1985).

While a number of congressional statutes play roles in forest policy, three are the most crucial. The most fundamental is the National Forest Management Act (NFMA) of 1976. NFMA changed both the substance and the procedures of forest policy through the creation of a new planning process and requirements for forest practice regulations. NFMA established a planning process in which the Forest Service is required to prepare a long-term integrated plan for each national forest.[3] Public participation, particularly by environmental groups, was dramatically expanded by the new planning process. NFMA required public participation in the "development, review, and revision" of forest plans (US Office of Technology Assessment 1992: Chapter 5). In implementing those requirements, the Forest Service created its own elaborate process of administrative appeals, giving the public the right to appeal decisions ranging from forest plans to specific timber sales (Bobertz and Fischman 1993).

The second major feature of NFMA was the requirement that it establish forest practice regulations for national forests. Unlike a number of other environmental statutes passed in the 1970s, NFMA did not create any major substantive restrictions on agency discretion. The fundamental task of balancing multiple uses was still left largely for the agency to handle. NFMA did require the Forest Service to create standards and guidelines for timber management and protection of other resources.

When it promulgated these regulations, the Forest Service imposed a number of restrictions on its own discretion (Wilkinson and Anderson 1987: 119). Perhaps the most famous standard was the language chosen to implement the statute's protection of wildlife. NFMA requires that forest planning "provide for diversity of plant and animal communities based on the suitability and capability of the specific land area in order to meet overall multiple-use objectives" (16 U.S.C. sec. 1604[g][3][B]). The implementing regulations transformed this general guideline into a stringent action-forcing requirement: "fish and wildlife habitat shall be managed to maintain viable populations of existing native and desired non-native vertebrate species in the planning area" (Wilkinson and Anderson 1987: 296). To the surprise of many Forest Service officials, this language became the centrepiece of the environmentalists' litigation strategy to stop logging in old-growth forests in the US Pacific Northwest to preserve the northern spotted owl and other vulnerable species.

In addition to NFMA, forest policy is directly affected by the National Environmental Policy Act (NEPA) of 1969. NEPA requires the filing of an environmental impact statement prior to the undertaking of any "major federal activity." Because timber management on federal lands is such an activity, NEPA impact assessment requirements have played a fundamental role in the old-growth dispute.

The third key statute is the Endangered Species Act of 1973, which the US Supreme Court has called "the most comprehensive legislation for the preservation of endangered species ever enacted by any nation" (*TVA v. Hill*, 437 U.S. 153, 180 [1978]). The Fish and Wildlife Service is authorized to "list" species as threatened or endangered. This classification must be based solely on biological information and cannot include economic considerations about the consequences of such a listing. Economic considerations can be incorporated, however, into the design of habitat management plans. The consequences of declaring a species to be threatened or endangered are immense. The act prohibits any federal agency from engaging in activities that "jeopardize the continued existence of any endangered or threatened species" or "result in the destruction or adverse modification of habitat which is determined ... to be critical" (16 U.S.C. sec. 1536[a][2]).

These statutes, and the regulations promulgated under their authority, transformed the regime governing federal forest policy. By creating a complex web of substantive and procedural requirements, the new rules limited the discretion of an implementing agency and granted extensive opportunities for public input and appeal. As a result, the new requirements forced the agency to pay more attention to the concerns of environmental groups. But they also created a procedural quagmire for the agency. A 1990 estimate was that the planning process costs $200 million annually, and virtually all of the plans have been subjected to administrative appeal, and many aspects of them have met with court challenges (*Economist,* 10 March 1990: 28).

As in many other areas of natural resource policy, courts have come to play a pivotal role in policy formation. In the 1960s, the Forest Service faced an average of one lawsuit per year. By the early 1970s, litigation intensified to an average of two dozen a year (Brizee 1975). By 1991, pending litigation before the agency consisted of ninety-four cases (US Office of Technology Assessment 1992: 100), and it stayed at about that level for the remainder of the decade.[4]

The BC Policy Regime: Executive-Centred Bargaining

In contrast to the United States, in Canada jurisdiction over forests belongs almost exclusively with the provinces. At the provincial level in British Columbia, the government owns virtually all (95 percent) of the forested land, and thus the BC Ministry of Forests is the key regulator. The province's forest land is managed through a complex system of "tenures." The two dominant forms of tenures are the area-based tree farm licences (TFLs) and the volume-based forest licences (FLs) organized into timber supply areas (TSAs). As of 1996, 58 percent of British Columbia's annual allowable cut was represented within thirty-seven TSAs, and 24 percent of the allowable cut was represented by thirty-three TFLs. A small business program makes up another 14 percent of the allowable harvest (Wilson, Wang, and Haley 1999).

In both TSAs and TFLs, the Ministry of Forests delegates certain forest management responsibilities to private timber companies in exchange for long-term guarantees of timber supply. The specific balance of responsibilities varies according to the form of tenure. Holders of TFLs tend to have much greater planning and management responsibility than companies operating in TSAs. In exchange, TFLs are granted for longer terms. Despite public ownership of the land, these tenure rights are taken seriously in BC law. Companies have the ability to keep members of the public away from certain areas if they are deemed to interfere with logging activities, and, if the government decides to set aside some of the land allocated in the licences, then it must compensate the licence holders. The price that the government charges for allowing harvesting of "Crown" timber – called "stumpage" – is determined by a complex administrative formula. The fact that stumpage prices are not more directly linked to market signals has made the BC system vulnerable to charges by American timber interests that it constitutes an unwarranted subsidy (Cashore 1997).

While pluralist legalism best characterizes the US environmental policy regime, Canadian environmental policy is better characterized as executive-centred bargaining. Rather than separating the legislative and executive branches, parliamentary systems combine them. The cabinet and the prime minister are selected from the party with the most seats in the legislature. This fact, combined with exceptionally strong party discipline, means that the executive (cabinet) of the majority party dominates policy making, and there is virtually no significant role played by the legislature independent of the cabinet. The absence of an independent legislature means that there are no incentives for the government to create the types of nondiscretionary duties that dominate the American policy landscape. Legislation generally provides broad grants of authority but almost never binds the "Crown" – as Canadians, reflecting their British roots, call the formal government – to perform any particular task.

As a result of this enabling structure of statutory law, policy tends to result from bargaining between the executive and the relevant societal interest groups. Prior to the late 1980s, environmental groups in Canada tended to play minor roles in the policy process, and most of the relevant bargaining was between government and business (Wilson 1998). Since then, however, the policy process has become more pluralistic, with the bargaining process expanded to include environmentalists and other relevant stakeholders (Hoberg 1993a).

Forest policy in British Columbia is guided by a regulatory framework that has been executive-centred and highly discretionary. Prior to the major reforms of the 1990s, BC forest management was governed by the Ministry of Forests Act and the Forest Act, both of which set out general standards for forest management that do little to constrain the discretion of the minister of forests or the chief forester. In recent years, other government departments have become increasingly relevant to forest management. The federal Department of Fisheries and Oceans has some jurisdiction because of the influence of forestry on salmon habitat, and the BC Ministry of Environment has become increasingly important because of both greater concern for the environmental impacts of forestry and its jurisdiction over parks. Because of the significance of forest policy to the province, the cabinet typically plays a large role in most major decisions, particularly those involving land use.

Whereas the US forest policy regime underwent fundamental change in the early to mid-1970s, the BC forest policy regime was relatively stable until the 1990s, when pressure for institutional reform escalated. The 1991 election of the social democratic New Democratic Party (NDP) brought an end to sixteen years of rule by the strongly pro-business Social Credit Party. The NDP was far more pro-environment, and the 1990s witnessed major changes in both institutions and policies. There have been two major pushes for institutional change. The first, initiated by the government, has been a highly innovative effort to use consensus-based negotiation to resolve land-use disputes. The second, initiated by the environmental community, has been an effort to import US-style legalism into BC forest policy. While it has less to do with institutional change, a strategy of appealing to international markets has also been adopted by environmentalists and has been relatively successful.

Consensus-Based Negotiations
The use of consensus-based multistakeholder consultations began in 1989 as the Social Credit Party struggled with the environmental controversies raging around it. The most comprehensive effort to employ "shared decision making" (as it has come to be called) was the creation of the Commission on Resources and the Environment (CORE), headed by Stephen Owen. CORE was charged with developing a comprehensive land-use plan for British Columbia, and it began by creating three regional processes, starting with Vancouver Island, "to resolve land use disputes by defining protected areas and providing greater certainty on lands available for integrated resource management." It was also charged with applying "mediation/facilitation dispute resolution techniques to land use issues" (Premier's Office 1992).

The major activities of CORE were three regional "roundtables" to develop land-use plans: Vancouver Island, Cariboo-Chilcotin, and Kootenay-Boundary. Despite extraordinarily intensive and time-consuming negotiations, not one of the roundtables was able to achieve consensus on the most important issue: where to draw lines on the map to allocate land to wilderness and logging. As a result, that key task was left to CORE staff members, who developed plans and then submitted them to the cabinet. In some cases, the cabinet made significant revisions to the CORE recommendations on which areas were to be protected and which were to be logged, but in all cases CORE plans were extremely influential in the government's decisions. In no case were there significant changes in the aggregate amount of land allocated to the three major categories.

Because of its failure to achieve consensus among stakeholders, the CORE process was not as successful as many of its proponents had hoped. However, it did change governance of the land-use process by institutionalizing environmental values in a new and effective way in the policy-making process.[5] While cabinet had the final say – and did make some substantial alterations – the plans developed by the environmentally oriented CORE staff were instrumental in shaping the final product. Aside from these institutional changes, the CORE process resulted in major victories for environmentalists, who achieved significant additions to protected areas in the province and advanced comprehensive land-use planning by contributing to the resolution of a number of divisive conflicts across the province.

CORE was quietly terminated by Premier Glen Clark shortly after he took over from Mike Harcourt in the spring of 1996, and its residual activities were incorporated into the interagency Land Use Coordination Office. The spirit of CORE lives on, however, in the so-called subregional Land and Resource Management Plans (LRMPs) being drafted around the province. Thus far, there are eighteen LRMPs, ten of which were completed by November 1999. Many of these LRMP processes have managed to achieve the consensus among stakeholders that eluded the CORE process (Wilson 1999: Chapter 11).

Importing Legalism

While these new consultation processes opened the way for previously marginalized interests, not everyone was satisfied with the direction of institutional change. In particular, a new environmental group was formed, modelled after the archetype of American environmental law firms, the Sierra Club Legal Defense Fund, right down to its name. The Sierra Legal Defence Fund (note spelling) was formed in 1990 with "an ambitious plan to protect Canada's environment through legal action" (Sierra Legal Defence Fund n.d.).[6]

Overall, environmentalists have not been successful in court to constrain the Ministry of Forests to produce more environmentally friendly decisions. Of the six major cases directly relevant to BC forest law, only two could be classified as successful, but one was subsequently overturned by another court, and the other was rendered moot by administrative response (Hoberg 1997b). The one major decision by the courts to constrain the discretion of the Ministry of Forests was actually a lawsuit brought by a major forestry company against a decision by the chief forester to reduce its allowable cut level. The two lawsuits most relevant to old-growth preservation are described briefly in a subsequent section.

The limited effect of this litigation is not surprising given the structure of the law that courts are asked to apply – it is enabling but almost never mandatory. The bread and butter of so-called action-forcing statutes in the United States – the requirement for certain administrative actions within a specific period of time – is simply unavailable to BC environmentalists. The few deadlines in the Forest Act can be easily circumvented by its remarkable Section 152, which gives blanket authority to the minister to "extend a time required for doing anything under this Act" (Forest Act, RSBC 1996, Chapter 157).

The litigation campaign – combined with other political pressures – has had some effect on legislative and procedural design, most of it beneficial to environmentalists. As a result of litigation, the process of determining allowable annual cuts has changed fundamentally. Rather than a quiet negotiation between the chief forester and the timber company, there is now a multistage process of analysis that explicitly calls for and incorporates public comment. When announcing the final decision, the chief forester publishes an extensive statement of rationale.

The Forest Practices Code (FPC)

The biggest change to BC forest policy has been the enactment of the Forest Practices Code, a legal codification of many of the diverse guidelines and regulations that were previously applied on a contractual basis to timber companies. The code has led to a significant formalization of the regulatory process, but it still reflects the powerful imprint of the macropolitical context. The new FPC came into effect in June 1995. Introducing the measure, the government claimed that it would produce "world class forest practices" and emphasized stiff enforcement penalties of up to $1 million a day (Hamilton and Baldrey 1994). Structurally, the code has three components: the Code Act enacted by the BC legislature, a series of binding regulations issued through Orders in Council, and a set of non-binding guidebooks meant to facilitate implementation of specific provisions. Functionally, the code also has three components: enforcement, forest practices, and planning.

The FPC created new regulations and guidelines for specific forest practices. For example, the regulations limit the size of a clear-cut – forty hectares in the coastal and southern interior regions and sixty hectares in the northern interior region. The code also created rules for stream protection, such as buffers around certain types of streams and guidelines for harvesting around other types. The code also introduced a new system for compliance and enforcement, including a dramatic increase in the maximum fine, from $2,000 to $1 million. In its public relations on the code, the government emphasized the stringent fines as an indication of its tough new approach to enforcement. A Compliance and Enforcement Branch was created in the Ministry of Forests, and the Code Act provided for a Forest Appeals Commission to hear complaints of licensees about enforcement actions.

At heart, the FPC is a framework for planning. "Higher-level plans" are essentially zoning decisions about which values are emphasized – they include resource management zones, landscape unit objectives, and sensitive area designations. These higher-level plans guide the development of "operational plans" oriented toward more specific aspects of forestry. Originally, the FPC provided for six operational plans: forest development plans, logging plans, silvicultural prescriptions, stand management prescriptions, five-year silvicultural plans, and access management plans. The central planning document was the Forest Development Plan (FDP), a five-year plan laying out all the proposed management activities for the area. The most important site-specific plan is the Silviculture Prescription (SP).

One of the most important features of the act was that, once higher-level plans were in place, operational plans had to be consistent with them. Some of the early environmental litigation attempted to get courts to require the ministry to approve various higher-level plans before it could approve site-specific cutting and road building. The new FPC significantly strengthens environmentalists' hands in this regard. It explicitly requires approval of forest development plans before road building can be authorized. Under the new rules, if the ministry behaves in the same way as it behaved in earlier cases, then it would clearly be in contravention of the law.

In addition to planning and forest practices, other aspects of the new regulatory framework created by the FPC were formalized. Perhaps the most significant was the creation, under the Forest Practices Code Act, of the Forest Appeals Commission and the Forest Practices Board. Under the BC Forest Act, appeals of ministry decisions are limited to licence holders. Environmentalists and other members of the public do not have the right to an administrative appeal, although they may be able to challenge ministry decisions in court under the Judicial Review Procedures Act. When the government introduced the code, it made a deliberate decision not to follow the US Forest Service route and make appeals available to industry and environmentalists equally. The Forest Appeals Commission is a quasi-judicial tribunal designed to adjudicate appeals by industry of a range of decisions by the government under the FPC.

Rather than locking environmental interests out of the appeal process entirely, however, the government followed the innovative path of creating a new administrative entity, the Forest Practices Board, to act as a public watchdog. The board does not have any legal or regulatory power, but it has the authority to conduct investigations into Ministry of Forests' implementation of the code and to audit corporate forest practices. It can also hear complaints from the public, including environmentalists, about ministry decisions under the code or corporate forest practices. The board does not have the authority to take action on a complaint, but it can investigate a complaint and make recommendations to the ministry. It can also launch an appeal to the Forest Appeals Commission on behalf of the public.

Thus, the FPC acts as a buffer between public complaints and the ministry – a bureaucratic floodgate holding back the rush of complaints demanding ministry attention that would likely occur if an American-style process were adopted. Although ministry decisions are still insulated from direct public challenge, the Forest Practices Board – through investigation and auditing – grants the public much more access to information about ministry decisions. Much of the impact of this reform clearly depends on the role that the board chooses to play.

The process has become more open and formal, but the procedural rights of environmentalists still pale compared with those of their

American counterparts. Environmentalists south of the border can directly appeal administrative decisions, and when they go to court they have far more formidable tools to use than do BC environmentalists. While far more formal in approach than the previous framework, the FPC still contains a provision that insulates the Ministry of Forests from challenges based on action deadlines. Section 165 authorizes the minister or a minister's delegate to "extend a time required to do anything under this Act, the regulations or the standards, other than a review or appeal of a determination or the time to commence a proceeding." Even after all these changes, the BC minister of forests, David Zirnhelt, proclaimed, "Don't forget the government can do anything" (cited in Beatty and Hamilton 1996). Of course, there is hyperbole in that statement, but it does reflect a certain Canadian mind-set that is anathema to Americans.

Going International

Because of the institutional structure of BC policy making, the strategic alternatives of environmentalists were relatively limited. The statutes as designed made litigation relatively ineffective. The provincial dominance over jurisdiction also made a strategy of nationalization difficult. There were appeals to the federal government to pressure the province into helping to create a national park, but not much effort was put into them because of the strong hand dealt to the province by the institutional framework. Lacking judicial or national alternatives, environmentalists embarked on an innovative international campaign that met with considerable success.

Moving beyond traditional strategies to influence governments by shaping public opinion, environmentalists began to take advantage of market forces to alter the incentives of corporations. Led by Greenpeace International, environmentalists began to target industrial consumers of BC forest products, initially in Europe and then in the United States, threatening to promote boycotts of their products if they did not stop purchasing BC forest products that environmentalists claimed were being produced in environmentally destructive ways (Bernstein and Cashore 1996; Stanbury and Vertinsky 1997). While few contracts were actually cancelled, the campaign succeeded brilliantly in giving BC forest companies and the province an economic interest in improving their environmental record. Despite increases in the costs of production, the companies came to accept that additional regulation was essential to maintaining market share.

Hypotheses about Institutional Effects

The two key differences in institutions between the two jurisdictions are legalism and federalism. The actual impact of these differences depends on the particular circumstances of the policy area. Perhaps the most important impact of legalism is that it gives environmentalists a great deal of control over a government's agenda. By petitioning agencies or mounting court challenges, environmental groups can pressure reluctant agencies to act. Without as many legal tools, Canadian environmental groups are far more reliant on mobilizing public opinion to pressure the government. But legalism does not grant tools just to environmental advocates. The doctrine of administrative law that can be used by the courts to force a reluctant agency to act can also be used to overturn pro-environment agency decisions challenged by business interests. The policy implications of the legalistic constraints on agency discretion thus depend on the balance between the legal tools held by environmentalists and those held by business. *Because of the pro-environment structure of American forest law, the greater legalism in the United States is an advantage for US environmentalists over their BC counterparts.*

What difference should the level of jurisdiction make? Depending on the structure of public opinion, changing the level of centralization may have profound consequences for the distribution of preferences on the issue.[7] If we consider a continuum of preferences ranging from pro-logging to pro-preservation, the question is how the balance of preferences among the population would change if the jurisdiction having control over the policy area changes. Many environmentalists in British Columbia, Washington, and Oregon are committed to preserving old-growth forests, but their interests are balanced by those of people who depend on forests for their livelihood and the overall contribution of the industry to the regional economy. The more important an industry is to a local or regional economy, the greater the electoral incentives are for policy makers to minimize the costs of production, including the costs of complying with regulations, in that industry (Harrison 1996; Mashaw and Rose-Ackerman 1984).

However, the further removed from the industry-dominated locality political control is, the fewer the electoral incentives to bow to the needs of that industry are. If we shift the jurisdiction to the national level, then the balance will likely shift toward preservation. Fewer people depend on old-growth harvesting for their livelihood, and, to the extent that they are concerned with the temperate rainforests at all, they are likely to be pro-preservation. These hypotheses are borne out by public opinion in the United States: people outside the region tend to have more pro-environment forestry stances than those within the region (see below). This change in the balance of preferences alters the political constraints on the relevant policy makers. Legislators from the region are faced with cross-cutting pressures and presumably try to search for some middle ground. In contrast, legislators outside the region, who have no vested political interest in the economic welfare of the forest-dependent area, may try to claim credit from environmental constituencies by saving ancient forests in the region. Therefore, *in the case of forest policy, there is good reason to believe that, in general, the more decentralized the jurisdiction, the less preservationist the policy outcome is likely to be.*

Court-Created Wilderness[8]
While substantial amounts of old growth were protected in national parks and wilderness areas created either by congressional statute or by administrative order, the modern battle over old-growth forests in the United States began in the late 1980s and was concentrated in the courts. Old-growth forests in the US Pacific Northwest did not emerge as a significant policy issue until late 1987.[9] There was an active and reasonably well-balanced forest policy subsystem in place prior to that date, but it was focused on designating alpine wilderness areas. A major shift occurred when the Sierra Club Legal Defense Fund (SCLDF) opened a new Seattle office in January 1987. The SCLDF launched a two-pronged legal strategy that has to be considered one of the most successful legal campaigns in the history of American environmental law.[10]

The first prong involved listing the spotted owl under the Endangered Species Act. In December 1987, the Fish and Wildlife Service (FWS) issued a decision that listing the spotted owl was unwarranted. The agency's own scientists had concluded the opposite, but the report was altered under the directions of Reagan political appointees (US General Accounting Office 1989). The SCLDF challenged the agency's decision in district court, and in November 1988 a federal district court vacated the FWS decision as "arbitrary and capricious" and remanded the issue to the agency for reconsideration.[11] In response, the FWS chose in June 1990 to list the owl as "threatened." While it turned out to have relatively little practical significance, this court ruling signalled the entry of the judicial branch into the old-growth controversy.

The second, and ultimately far more important, prong of the legal strategy was the series of legal challenges to the Forest Service efforts to comply with the requirements of NEPA and especially the NFMA in the district court in Seattle. In December 1988, the Forest Service finalized its supplemental environmental impact statement on the spotted owl and issued new regional guidelines for its protection. The SCLDF sued, and in March 1989 Judge William Dwyer, an appointee of Ronald Reagan, ruled that the plan was inadequate and issued his first injunction on timber sales in Washington and Oregon. This injunction, the first of many, was a pivotal event in the history of US Pacific Northwest forest policy because it shifted who benefited from the status quo. Now, for affected timber sales to go forward, the Forest Service had to comply with the judge's strict interpretation of the law or Congress had to take specific action to change the law as it applied in this case. Success in the judicial arena gave environmentalists new power in the executive and legislative arenas.

The Pacific Northwest delegation to Congress sought to regain control over the issue by attaching riders to appropriations bills exempting relevant logging activities from lawsuits.[12] The most prominent effort was Section 318, which, among other things, exempted both BLM and USFS timber sales from ongoing litigation.[13] In response to this setback, environmentalists revamped their strategy, fighting fire with fire. They reoriented litigation to focus on the constitutionality of Section 318, claiming that by attempting to decide the outcome of particular court cases Congress had violated the separation of powers.

Environmentalists also reconsidered their entire political approach, recognizing that, as long as old-growth forests were considered a regional issue, they would continue to lose in Congress. According to Andy Kerr of the Oregon Natural Resources Council, "expecting the Northwest congressional delegation to be rational about ending the cutting of ancient forests in the late 1980s is like expecting the delegation from the American south to deal rationally with ending segregation in the late 1950s" (personal interview). Environmentalists understood that in order to succeed politically they would have to nationalize the issue. Public opinion surveys show that there are significant differences between national and regional publics on these issues, with the national public being consistently more pro-environment (Steel, List, and Shindler 1992; Timber Industry Labor-Management Committee 1993).

Timing for nationalization of the old-growth debate could not have been better, as the environment more generally was gaining extraordinary salience nationwide. Emphasis was placed on the fact that the remaining old growth was virtually all in national forests, owned equally by all citizens of the United States. Feature stories appeared in the *New Yorker* and *National Geographic*, network news ran stories of activists sitting in trees in protest, and the issue reached the pinnacle of media exposure when the spotted owl made the cover of *Time* on 25 June 1990.

This successful campaign in the arena of public opinion was supplemented by national interest group mobilization. Groups not only sought to convince lawmakers outside the US Pacific Northwest that they had an electoral incentive to take an interest in the issue but also launched a more targeted political campaign to delegitimize the strategy of using appropriations riders to exempt Pacific Northwest forests from the application of environmental statutes (Sher and Hunting 1991: 487-90).

This amended environmental strategy was extraordinarily successful. In September 1990, the Ninth Circuit Court of Appeals took the dramatic step of striking down key parts of Section 318 as unconstitutional, on the ground that Congress directed a particular decision in pending litigation without amending the statutes used as the basis for litigation.[14] This decision was a stunning blow to timber interests and their allies in the Pacific Northwest congressional delegation because it invalidated their most effective means of insulating timber sales from environmental litigation.

The ruling turned out to be temporary, however, because in March 1992 the Supreme Court overruled the Ninth Circuit in a unanimous decision.[15] However, by that time the use of appropriations riders by regional legislators to reshape forest policy had been delegitimized by the concurrent political campaign by environmentalists to nationalize the issue. Legislators outside the region began to take an interest in the issue, and authorizing committees, whose statutes were being quietly rewritten, began to reassert their jurisdictional interests in the issue.

The focus of the process returned to efforts by the Forest Service and associated agencies to develop a plan for the protection of the spotted owl that could win judicial approval. A haphazard plan put together by the hostile Bush administration was challenged in court, and Judge Dwyer again ruled in favour of environmentalists, chastising the government for "a deliberate and systematic refusal ... to comply with the laws protecting wildlife."[16] The relevant law was the requirement in the regulations promulgated under NFMA, discussed earlier, for the maintenance of viable populations of wildlife. Dwyer ordered the Forest Service to develop "revised standards and guidelines to ensure the northern spotted owl's viability" by March 1992 and enjoined timber sales until it did so.

The Forest Service developed another plan, this time following proper procedures. In March 1992, it adopted the new plan based on the prestigious "Thomas report" setting aside about 8 million acres of old-growth forest for spotted owl habitat. Naturally, environmentalists sued again. In late May 1992, Judge Dwyer rejected the Forest Service's attempt to adopt the Thomas report as its spotted owl plan. The most striking part of the decision was his ruling that the plan was flawed because it did not adequately address issues related to species other than the spotted owl.[17] Continuing the pattern of previous cases, Dwyer imposed an injunction on timber sales until a satisfactory plan was put in place.

The decision stunned the Forest Service. Not only was the Thomas Plan, a state-of-the-art scientific document in 1990, ruled inadequate, but also the whole objective of the process was redefined by judicial order. The scope of the issue was significantly enlarged beyond one medium-sized owl to an entire ecosystem. A far more sophisticated analytical process was necessary to address this larger problem, and as a result the emphasis shifted from protection of particular species to management of an entire ecosystem.

Meanwhile, the politics of the executive branch was transformed with Clinton's election. Pro-timber officials were replaced by pro-environment ones, most prominent among them Interior Secretary Bruce Babbitt, previously the president of the League of Conservation Voters, and Chief Forester Jack Ward Thomas, the first biologist ever to hold the position. Rather than facing intense pressure from political superiors to water down their proposals to protect wildlife, the Forest Service and the Fish and Wildlife Service now confronted pressures to expand protection.

Making good on a campaign promise, the administration held a "forest summit" on 2 April 1993 in Portland. In a remarkable display of administrative commitment to problem solving, the president, vice-president, and six cabinet officials spent an entire day around a table listening to short speeches on one regional issue. In his closing remarks, Clinton committed his administration to the development of a plan that would be "scientifically sound, ecologically credible, and legally responsible" (Pryne and Matassa 1993). The process consisted of three working groups dominated by representatives of the relevant agencies: ecosystem management assessment, labour and community assistance, and agency coordination.

President Clinton announced his forest plan on 1 July 1993. The plan called for an annual harvest level of 1.2 BBF, which the scientific work group concluded was the maximum cut permissible under current law. In addition, the plan provided for extensive reserves for spotted owl protection and dramatically expanded riparian reserves for the protection of fish habitat. The scientific team assessed the plan's impact on the viability of over 1,000 species. Of the eighty-two vertebrate species analyzed, the plan was expected to provide an 80 percent likelihood of the maintenance of viable populations of all but three species of salamanders. In total, the plan set aside 80 percent of remaining old-growth forests. In an attempt to ease the pain in the region, the plan also provided for a massive $1.2 billion economic assistance package.

The compromise was bitterly attacked from all sides.[18] Industry and labour groups claimed that the dramatically reduced cuts would devastate timber-dependent rural communities. Environmentalists harshly criticized the size of the cut and especially the nature of the old-growth reserves, because some logging would be allowed for fire or insect salvage and some thinning of second-growth stands would be permitted to promote old-growth characteristics.

While environmentalists did their utmost to act as outraged as the timber industry and loggers, they had in fact achieved a remarkable victory. To put Clinton's plan in the proper perspective, one need only go back to 1989. During the debate over Section 318, environmentalists proposed an allowable cut level of 4.8 BBF per year. A harvest level that they were willing to accept in 1989 was four times higher than the level that they considered outrageously high in 1993. This shift by a factor of four in the harvest level indicates the dramatic redistribution of power achieved in this area by four years of effective lobbying in Congress, a successful public relations campaign to polish and nationalize the issue, and especially a brilliant litigation campaign.

Environmentalists were not satisfied, however, and once the plan was finalized in April 1994 they challenged it in court again. This time industry challenged the decision as well, arguing that the process used to develop the plan violated the Federal Advisory Committee Act. Judge Dwyer finally had a plan that he was willing to accept, and he upheld the Clinton forest plan in December 1994, brushing aside criticisms from both sides.[19]

While Dwyer's decision appeared to bring finality to the case, the political issue of old-growth forests reemerged with a bang after the dramatic Republican takeover of Congress in the 1994 elections. Committee leadership changed from strong environmental supporters to some of the Pacific Northwest region's most prominent supporters of the timber industry. Rolling back environmental regulations was a core component of the Contract with America; however, just as the Reagan administration had done before them, the Gingrich Republicans misinterpreted their mandate as including a desire to cut back on environmental protection. While this effort was largely unsuccessful, Republicans did sneak through one change with significant implications for old-growth preservation. This rider had three sections. The first involved easing environmental restrictions on "salvage logging," the logging of trees damaged by fire or disease. At first, this section raised the most concern for environmentalists, but a more careful reading of the other two sections of the timber rider intensified their concerns. One measure accelerated timber sales allowed under the Clinton forest plan by declaring them to be in compliance with other relevant laws and by insulating them from past or pending lawsuits. The third and more confusing measure ordered the Forest Service to award timber sales "subject to Section 318 of Public Law 101-121," the 1990 rider discussed above. Environmentalists and the Clinton administration were both surprised when an industry lawsuit achieved an expansive reading of this section.

Most of the political fight over forests in 1995 and 1996 was over this rider (Kriz 1996). Despite all the sound and the fury, the rider expired at the end of 1996, and industry and its supporters in Congress appear to have little inclination to rejuvenate the strategy. The rider did result in increased logging, some of it in old-growth forests, but industry and environmentalists both agree that the total amount is extremely small – less than 1 percent of the remaining old-growth forest in the region. There now seems to be a renewed sense of closure to the issue.

This case clearly shows how the particular institutions of the US forest policy regime affected both the strategy and the resources of environmental groups and contributed to their success. The environmental strategy in this case can be boiled down to two tactics: nationalization and judicialization. The victory would not have been possible had the issue continued to be constructed in regional terms, as forest policy traditionally had been. Nationalization was a viable strategy only because of the clear federal jurisdiction over old-growth forests. Judicialization was made possible by the peculiar feature of legalism so pervasive in environmental policy. US Pacific Northwest old-growth protection has probably become the most extreme case of judicial intervention into environmental policy making. From the time of his first injunction in 1989 to his approval of the Clinton forest plan in late 1994, Judge Dwyer essentially managed Region 6 of the US Forest Service. Environmentalists would respond that Dwyer was merely enforcing the law, and they have a point. The regulations promulgated to implement NFMA "diversity" requirements elevated the status of species protection in the agency's multiple-use equations and forced the agency into unexpectedly preservationist decisions.

Cabinet- and Committee-Created Wilderness

Whereas old-growth protection in the United States has been dominated by the federal government and the courts, in Canada it has been dominated by the BC provincial government and a pluralistic, executive-centred bargaining process. The first major attempt to use consensus-based negotiation to resolve a major land-use dispute in the province was Clayoquot Sound, a jewel of wilderness on the west coast of Vancouver Island. Initiated in 1989 under the Social Credit government, the outcome was not promising for shared decision making, because two separate attempts to reach compromise failed, both over where logging should occur while the talks were going on. With the failure of the innovative process, the locus of policy making shifted to an internal debate within cabinet. In April 1993, the government announced its decision to allow logging in two-thirds of Clayoquot Sound (British Columbia 1993). The decision provoked the outrage of environmentalists and resulted in a massive civil disobedience campaign, which led to over 800 arrests and a great deal of unfavourable international attention to BC forestry. The decision increased the protected area in the sound from 15 percent to 33 percent.

The second effort to employ shared decision making was a committee, set up in late 1989, designed to establish a strategy for the preservation of old-growth forests. This Old-Growth Strategy Committee made its recommendations to cabinet in 1992, and it was subsequently subsumed under the rubric of the more comprehensive wilderness planning process, called the Protected Areas Strategy, established in May 1992.

The Protected Areas Strategy became the umbrella initiative in wilderness protection in the province. The core of the program was a commitment by the province to double its protected areas to 12 percent of its total area by the year 2000. This 12 percent figure, derived from the recommendations of the famous Brundtland Commission report (World Commission on Environment and Development 1987) guided land-use discussions throughout the province. For my comparative purposes, the most important consequence of this policy is that it has served as a de facto ceiling on wilderness protection across the province. "Special" areas such as Clayoquot Sound have received proportionately greater protection, but the average across the province will not exceed 12 percent without major political change in the province. The strategy is designed to achieve representativeness of the province's "ecosections," but there is no requirement that the 12 percent target should be applied to each ecosection.

A third major effort to use shared decision making to resolve disputes over coastal old-growth preservation was the Vancouver Island roundtable of the Commission on Resources and the Environment (CORE). As in Clayoquot Sound, the stakeholders could not come to consensus about land-use allocation. As a result, CORE staff presented a plan to cabinet. The proposal provoked a massive demonstration by loggers and their supporters. Approximately 15,000 demonstrators crowded the lawn in front of the legislative building and shouted down Premier Harcourt. When the government announced its decision on the plan in June 1994, it bought thirty minutes of television time to explain the decision to the province. The plan increased protected areas from 10.3 percent of the region to 13 percent.

A number of other major decisions about protected areas have been made. In the summer of 1994, a large area along the north coast of the province – the Kitlope – was set aside. This decision resulted from quiet negotiations among four major parties: the forest company with cutting rights in the area, First Nations in the region, an American environmental group called Ecotrust, and the government (Gill 1994). In 1996, a multistakeholder panel achieved resolution in the hotly contested Lower Mainland (the southwest corner of the province), increasing protected areas from 10.5 percent to 13 percent. Most of the remaining areas are being negotiated as part of the subregional Land and Resource Management Plans (LRMPs). As noted above, these bodies are attempting to resolve disputes through consensus-based negotiation.

Cabinet and committee processes have thus dominated the old-growth preservation process in British Columbia. The BC environmental litigation campaign has had no substantive effect on old-growth preservation. Two cases provide clear contrasts to the success of the US strategy. In the marbled murrelet case in 1991, environmental groups tried to get the courts to rule that a federal environmental assessment was required for old-growth harvesting of areas believed to be nesting habitat for the seabird. Their argument was that the murrelet was a migratory bird, that the federal government had jurisdiction over migratory birds, and therefore, under evolving environmental assessment jurisprudence, that a federal assessment of logging activities was required.[20] The federal trial court disagreed and dismissed the case without any written reasons.

The only lawsuit to involve the northern spotted owl was a case about annual allowable cut (AAC) determinations. Environmentalists challenged the chief forester's failure to include the effects of potential restrictions on harvesting resulting from the finalization of the proposed recovery plan for the spotted owl. The case involved determination of the AACs for the Soo and Fraser Timber Supply Areas in 1995. A recovery plan for the spotted owl had been in the works since 1990. In June 1995, the government announced a general strategy to guide development of a spotted owl management plan, but it had not yet come to any conclusion. The recovery plan was being worked out between the relevant agencies and cabinet, with no court involvement and only general consultations with environmentalists, who argued that the chief forester acted unlawfully by failing to account for expected withdrawals of harvestable timber from the plan. The court agreed with the chief forester that the spotted owl recovery plan was a land-use decision under the jurisdiction of cabinet and that it was therefore reasonable to exclude it from his determinations.

In its opinion, the court took pains to explain operation of the BC political system. Reconciliation of the conflicting values of species protection and timber harvesting "involves land-use decisions which are properly addressed by government within the political arena rather than by the Chief Forester within his administrative mandate." Even if they were within the chief forester's mandate, "The court has no role in fixing the AAC. That discretion is given to the Chief Forester alone."[21] This is a far cry from the view of Judge Dwyer, who saw fit to impose injunction after injunction until he was satisfied that the government's spotted owl plan met the legal standards. The differences are not simply a reflection of different judicial traditions. They have more to do with the law that the judge is asked to apply. There are statutes designed to protect endangered species in Canada and BC, but they have no nondiscretionary duties in them. There is nothing comparable to the NFMA's viability requirements in the United States.

The BC government finally announced its spotted owl decision on 9 May 1997. The plan did not create any new protected areas; instead, it relied on more established parks and the new protected areas of the Lower Mainland Protected Areas Strategy. These protected areas make up 159,000 of the 1,105,000 hectares (14 percent) of forested habitat in the spotted owl range.[22] In addition, the plan created 204,000 hectares of "special resource management zones" in which a minimum of 67 percent of suitable spotted owl habitat must be maintained, defined as "forest that is more than 100 years old." The decision was not presented in a way that makes for straightforward comparisons to the US decision (e.g., what percent of old-growth forests in the area is being protected). The Ministry of Forests estimates that there are 372,000 hectares of old-growth forest outside parks in the area. Assuming that all the forested area in the parks is old growth, parks would protect 30 percent of old growth in the spotted owl range. No estimate has yet been made of the additional old-growth protection that would result from the special resource management zones, but this category has certainly received far less protection than the late-successional reserves created by the US plan. One clear indication of the different approaches is the impact on allowable cut levels. The BC plan is estimated to reduce allowable cuts in the region by 3.5 to 5 percent (British Columbia 1997). The US decision reduced allowable cuts by about 75 percent.

Thus, the process of old-growth preservation in British Columbia has been dominated by cabinet decision making following multistakeholder consultations. These consultations have had some success in achieving consensus, but more often than not the difficult allocative decisions have had to be made by cabinet. The process is far more pluralistic than it was prior to 1990 in that environmentalists have a permanent place at the table. But the executive-centred and federally decentralized form of decision making limits the strategies and the political resources available to advocates of greater preservation. Because of the design of statutes, however, the litigation campaign has had limited effects. The dominance of provincial jurisdiction prevented the type of nationalization strategy so successfully pursued by American environmentalists. BC environmentalists adapted to the absence of these political resources and adopted instead a strategy designed to influence BC policy through international markets. While it is impossible to estimate the influence of that international campaign, there is little doubt that it has significantly strengthened the political resources of environmentalists.

Resulting Old-Growth Protection

The United States has protected substantially more of its remaining old-growth forests than has British Columbia. The US government estimates that there are 8,551,000 acres (3,461,000 hectares) of old-growth forests on federal lands. Prior to the litigation campaign, 2,901,000 acres (1,174,000 hectares) (34 percent) was set aside in national parks and congressionally or administratively designated wilderness areas. The Clinton forest plan led to the additional protection of 3,923,000 acres (1,588,000 hectares) (47 percent). Thus, 57 percent of the area now protected resulted from the recent spotted owl controversy.[23] *This addition brings the area of protected old growth as a percentage of total remaining old growth to 80 percent in the US Pacific Northwest.*

In British Columbia, the amount protected is far less. According to estimates by Ministry of Forests staff, there are 26.2 million hectares of remaining old growth provincewide, 2.7 million, or 10 percent, of which have been protected. Of that protected area, 2.2 million hectares were protected prior to 1992. If we look at just the coastal area, which is more directly comparable to the US figures, then the percentage is less. Along the coast, there are 6,688,000 hectares of remaining old growth.[24] Of this total, 569,000 hectares, or 8.5 percent, have been protected to date. And 31 percent of that protected area has been set aside since the Protected Areas Strategy began in 1992 (MacKinnon and Vold 1998). The Protected Areas Strategy is still designed to meet its 12 percent target provincewide. Absent fundamental political change in the province, that *12 percent is likely to be an upper limit for the portion of coastal old growth set aside as wilderness in British Columbia.*

Conclusions

This analysis has shown that (1) forest policy is made in profoundly different ways in British Columbia and the United States, (2) the protection of old-growth forests in each jurisdiction has reflected these broader patterns, and (3) the United States has protected comparatively more old-growth forest than has British Columbia. Going beyond these associations to establishing a causal link between the different institutions and the different policy outcomes is far more complicated. There are several other major categories of variables that could account for the outcome. This section describes some of those variables and presents counterarguments to support the case for the independent influence of institutions.

The policy problems in the two jurisdictions differ significantly in terms of both the amount of remaining old growth and the scope of the endangered species problem. The United States began active harvesting earlier than British Columbia and has already harvested virtually all (85-90 percent) of its old-growth forests. As a result, it can support a strong timber industry on second-growth forests. In British Columbia, there is very little second growth ready to harvest, so the forest economy is far more dependent on old growth. This argument can be countered by looking at an area of the United States where old-growth coastal forests are still relatively abundant, the Tongass National Forest in southeastern Alaska. Of the 5.5 million acres of old growth in Alaska, 5.1 million acres remained in 1988. Only 6 percent has been logged. If the developmental view is accurate, then old-growth preservation in Alaska should be closer to British Columbia than to Oregon and Washington. Yet of the 5.1 million remaining acres, 2.2 million, or 43 percent, have been set aside as wilderness by Congress (Hoberg 1993b). The fact that so much more old growth has been set aside in Alaska, an area where the resource is still extremely abundant, than in British Columbia suggests that the explanatory power of this "developmental view" is limited.

Another significant difference in the problem is the scope of the endangered species issue. An argument could be made that US environmentalists were so successful because their legal hook covered such an expansive range: in the United States, the spotted owl exists throughout the entire range of temperate rainforest, whereas in British Columbia it exists only in the southwest portion of the mainland. But we can see from the different types of decisions made that this argument has limited force. British Columbia's spotted owl decision was carefully limited to minimize the impact on timber harvesting, and it was not guided by the overwhelming concern for species viability that occurred south of the border. The nature of US law – and the tools that it gave to environmentalists to force the government to act – thus comprise an important independent explanation.

Another important counterargument is that the difference is driven by different levels of economic dependence on forest sectors. In 1991, 5.6 percent of the BC workforce was directly employed in the forest industry, whereas the rate in Oregon and Washington was only 3.3 percent (Hoberg 1993b). This difference in economic dependence might explain why the forces resisting greater old-growth protection have been more powerful in British Columbia.[25] This argument is less easily discredited. But the timber industry is also formidable south of the border, and it is important to consider what would have happened there if not for the particular institutional features of forest policy. Significant restrictions on harvesting in old growth were not applied until environmentalists brought their lawsuit to bear. Conceivably, Congress or the administration would have acted by now to preserve more old growth. (Indeed, it is Congress that has protected so much old growth in Alaska.) But given the power of the regional delegation, and other pro-industry interests, to use one of the multiple veto points in the system to block such changes, it is hard to imagine a situation in which so much old growth could have been protected without the political resource of legal injunctions.

Institutional differences are not the only reason why the United States has protected more old growth than British Columbia, but clearly they have had some independent explanatory effect. Pluralist legalism and centralized federalism do not automatically empower environmental groups; however, in the particular circumstances of forest law and old-growth protection, American institutions have been a tremendous resource for environmentalists.

Notes

1 For other efforts to compare forest policy in the two jurisdictions, see Cashore (1995) and Leman (1987).
2 The literature on the effects of institutions is vast. For the historical tradition of institutionalism, see Evans, Rueschemeyer, and Skocpol (1985) and Steinmo, Thelen, and Longstreth (1992); for the organizational tradition, see March and Olsen (1989); and for the rational choice tradition, see Shepsle (1986).
3 This planning requirement was originally established by the Forest and Rangeland Renewable Resources Planning Act of 1974, but the act was fundamentally revised by the NFMA before it could be fully implemented.
4 According to the USDA Office of General Counsel (1999), the Forest Service faced ninety-five pending cases in December 1999.
5 For an alternative assessment, see the chapter by Mae Burrows in this volume.
6 The group got off the ground with a grant from the Law Foundation of British Columbia, and, in addition to individual members, it has received support from the US Sierra Club Legal Defense Fund and the US Bullitt Foundation, which supports environmental groups. The forest industry found the SCLDF litigation strategy sufficiently offensive that it wrote to the Law Foundation requesting that it discontinue funding. The industry group charged, among other complaints, that "the SCLDF has a closer relationship with United States interests than it does with the interests of the general public in British Columbia" (BC Forest Industry Land Use Task Force 1992).
7 For an overview of the spatial theory of politics, see Strom (1990).
8 This section relies heavily on Hoberg (1997a).
9 A thorough account of the spotted owl controversy, focusing on the administrative and interest group aspects, can be found in Yaffee (1994).
10 There was a third prong to the SCLDF legal strategy involving the lands managed by the Bureau of Land Management in Oregon, but I will not address it here.
11 *Northern Spotted Owl* v. *Hodel*, 716 F.Supp. 479 (W.D.Wa. 1988).
12 As Sher and Hunting (1991) document, appropriations riders were an attractive and proven mechanism for members of Congress to override the implementation of environmental statutes when they had harsh local consequences.
13 Public Law No. 101-121, sec. 318, 103 Stat. 701, 745-50 (1989). For a detailed discussion of the events surrounding the first "timber summit" and the enactment of Section 318, see Balmer (1990) and Johnston and Krupin (1991).
14 *Seattle Audubon Society* v. *Robertson*, 914 F.2d 1311 (9th Cir. 1990).
15 *Robertson* v. *Seattle Audubon Society*, 112 S.Ct. 1407 (1992). The Supreme Court ruled that appropriations riders did amend the relevant statutes and therefore did not violate the separation of powers.
16 *Seattle Audubon Society* v. *Evans*, 771 F.Supp. 1081 (W.D. Wash. 1991). The decision was upheld on appeal, *Seattle Audubon Society* v. *Evans*, 952 F.2d 297 (9th Cir. 1991).
17 *Seattle Audubon Society* v. *Moseley*, U.S. District Court, Western District of Washington, C92-479WD, 28 May 1992, 2 July 1992.
18 See, for instance, the extensive coverage in the *Seattle Times*, the *Seattle Post-Intelligencer*, and the *Portland Oregonian* on 2 July 1993.
19 *Seattle Audubon Society et al.* v. *Lyons*, No. C92-479WD, Order on Motions for Summary Judgment re 1994 Forest Plan, U.S. District Court, Western District of Washington, 21 December 1994. In February 1995, the main environmental coalition represented by the Sierra Legal Defense Fund announced that it would not appeal Dwyer's decision. Both industry and more extreme environmental groups appealed, however, and Dwyer's decision was upheld by the 9th Circuit in April 1996, *Seattle Audubon Society* v. *Moseley*, 80 F.3d 1401 (9th Cir. 1996).
20 See *Western Canada Wilderness Committee* v. *Minister of Environment and Her Majesty in Right of British Columbia*, Action No. T-2913-90, Federal Trial Court, 21 December 1990.
21 *Western Canada Wilderness Committee* v. *Chief Forester* (1996), 62 AC.W.S (3d) 779 (B.C.S.C.).

22 The amount of protected areas comes from British Columbia (1997). The figure for total harvested area comes from personal communication with Myles Mana, Vancouver Region, Ministry of Forests, 27 May 1997.

23 These figures are from the US Department of Agriculture and the Department of Interior (1994, 3&4-41). The amounts protected prior to the litigation campaign were "congressionally reserved areas" and "administratively withdrawn areas." The amounts protected by the Clinton forest plan were the "late-successional reserves" and the "riparian reserves."

24 These amounts include both the coastal western hemlock zone and the mountain hemlock zone.

25 The different stakes in old-growth harvesting are magnified by the particular political economy of the timber industry in the US Pacific Northwest. Large timber companies in the region, such as Weyerhaeuser, rely almost exclusively on their own tree farms and do not depend on harvesting federally controlled old growth to sustain their operations. For the most part, only small local operations depend on federal timber sales of old growth. Consequently, the effect of significant reduction in logging on federal lands may be windfall profits for the largest forest companies that benefit from the price effects of the restriction on timber supply to the region as a whole (Dietrich 1992: 186; Stevens 1993).

References

Arnold, Douglas. 1979. *Congress and the Bureaucracy*. New Haven: Yale University Press.

Balmer, Donald G. 1990. "United States Federal Policy on Old-Growth Forests in Its Institutional Setting." *Northwest Environmental Law Journal* 6: 331-60.

BC Forest Industry Land Use Task Force. 1992. Letter to the Board of Governors, the Law Foundation of British Columbia. 15 January.

Beatty, Jim, and Gordon Hamilton. 1996. "NDP Grab Hints Deficit Near $1 Billion." *Vancouver Sun* 13 September: A1.

Bernstein, Steven, and Ben Cashore. 1996. "The Internationalization of Domestic Policy Making: The Case of Eco-Forestry in British Columbia." Paper prepared for delivery at the Annual Meeting of the Canadian Political Science Association, Brock University, St. Catharines, ON, 2-4 June.

Bobertz, Bradley, and Robert Fischman. 1993. "Administrative Appeal Reform: The Case of the Forest Service." *University of Colorado Law Review* 64 (2): 371-456.

British Columbia. 1993. *Clayoquot Sound Land Use Decision*. Victoria: Government of British Columbia.

—. 1997. *Spotted Owl Management Plan: Summary Report*. Victoria: Crown Publishing.

Brizee, Clarence. 1975. "Judicial Review of Forest Service Land Management Decisions." *Journal of Forestry* 73: 424-25, 516-19.

Caldiera, Gregory A. 1991. "Courts and Public Opinion." In John Gates and Charles Johnson (eds.), *The American Courts*. 303-34. Washington, DC: Congressional Quarterly.

Cashore, Benjamin. 1995. "Comparing the Eco-Forest Policy Regimes of British Columbia and the U.S. Pacific Northwest." Paper prepared for delivery at the Annual Meeting of the Canadian Political Science Association, Montreal.

—. 1997. "Flights of the Phoenix: Explaining the Durability of the Canada-US Softwood Lumber Dispute." *Canadian-American Public Policy* 32: 1-58.

—. 1999. "US Pacific Northwest." In Bill Wilson et al. (eds.), *Forest Policy: International Case Studies*. 47-80. Oxon, UK: CABI Publishing.

Dietrich, William. 1992. *The Final Forest*. New York: Simon and Schuster.

Evans, Peter, Dietrich Rueschemeyer, and Theda Skocpol (eds.). 1985. *Bringing the State Back In*. Cambridge, UK: Cambridge University Press.

Gill, Ian. 1994. "The Lesson of the Kitlope." *Georgia Straight* 19-26 August: 7-11.

Hamilton, Gordon, and Keith Baldrey. 1994. "Forest Critics Seek Meat in NDP's Plan." *Vancouver Sun* 17 May: A1.

Harrison, Kathryn. 1996. *Passing the Buck: Federalism and Canadian Environmental Policy*. Vancouver: UBC Press.

Hoberg, George. 1992. *Pluralism by Design: Environmental Policy and the American Regulatory State.* New York: Praeger.

—. 1993a. "Environmental Policy: Alternative Styles." In Michael Atkinson (ed.), *Governing Canada: State Institutions and Public Policy.* 307-42. Toronto: HBJ-Holt.

—. 1993b. "Regulating Forestry: A Comparison of Institutions and Policies in BC and the US Pacific Northwest." FEPA Working Paper 185. Forest Economics and Policy Analysis Unit, Vancouver.

—. 1997a. "From Localism to Legalism: The Transformation of Federal Forest Policy." In Charles Davis (ed.), *Western Public Lands and Environmental Politics.* 47-73. Boulder, CO: Westview Press.

—. 1997b. "The Crisis of Governance in British Columbia Forestry." Unpublished paper.

Johnston, Bryan, and Paul Krupin. 1991. "The 1989 Pacific Northwest Timber Compromise: An Environmental Dispute Resolution Case Study of a Successful Battle that May Have Lost the War." *Willamette Law Review* 27: 613-43.

Kriz, Margaret. 1996. "Timber!" *National Journal* 3 February: 252-57.

Leman, Christopher K. 1987. "A Forest of Institutions: Patterns of Choice on North American Timberlands." In Elliot J. Feldman and Michael A. Goldberg (eds.), *Land Rites and Wrongs: The Management, Regulation, and Use of Land in Canada and the United States.* 149-200. Cambridge, MA: Lincoln Institute of Land Policy.

MacKinnon, Andy, and Terje Vold. 1998. "Old-Growth Forests in British Columbia II: Inventory." *Natural Areas Journal* 18(4): 309-18.

March, James, and Johan Olsen. 1989. *Rediscovering Institutions.* New York: Free Press.

Mashaw, Jerry L., and Susan Rose-Ackerman. 1984. "Federalism and Regulation." In George C. Eads and Michael Fix (eds.), *The Reagan Regulatory Strategy.* 111-45. Washington, DC: Urban Institute Press.

Moe, Terry. 1985. "The Politicized Presidency." In John Chubb and Paul Peterson (eds.), *The New Directions in American Politics.* 235-71. Washington, DC: Brookings Institution.

Niskanen, William. 1971. *Bureaucracy and Representative Government.* Chicago: Aldine-Atherton.

Premier's Office. 1992. "Harcourt Unveils Comprehensive Land Use Initiative." Press release, 21 January.

Pryne, Eric, and Mark Matassa. 1993. "Clinton Not in Favor of Changing Environment Laws or Halting Suits." *Seattle Times* 3 April: A1.

Shepsle, Kenneth. 1986. "The Positive Theory of Legislative Institutions: An Enrichment of Social Choice and Spatial Models." *Public Choice* 50: 135-78.

Sher, Victor, and Carol Sue Hunting. 1991. "Eroding the Landscape, Eroding the Laws: Congressional Exemptions from Judicial Review of Environmental Laws." *Harvard Environmental Law Review* 15: 435-91.

Sierra Legal Defence Fund. N.d. *Sierra Legal Defence Fund: Law Firm for the Environment.* Brochure.

Stanbury, W.T., and Ilan Vertinsky. 1997. "The Use of the Boycott Tactic in Conflicts over Forestry Issues: The Case of Clayoquot Sound." *Commonwealth Forestry Review* 76: 18-24.

Steel, Brent, Peter List, and Bruce Shindler. 1992. "Oregon State University Survey of Natural Resource and Forestry Issues." Corvallis, OR, 15 January.

Steinmo, Sven, Kathleen Thelen, and Frank Longstreth (eds.). 1992. *Structuring Politics: Historical Institutionalism in Comparative Analysis.* Cambridge, UK: Cambridge University Press.

Steinmo, Sven, and Jon Watts. 1995. "It's the Institutions, Stupid! Why Comprehensive National Health Insurance Always Fails in America." *Journal of Health Politics, Policy, and Law* 20: 329-72.

Stevens, John H. 1993. "Uncertainty Is Clouding Timber Firm's Rosy Outlook." *Seattle Times* 12 April: C1.

Strom, Gerald S. 1990. *The Logic of Lawmaking: A Spatial Theory Approach.* Baltimore: Johns Hopkins University Press.

Timber Industry Labor-Management Committee. 1993. "The Endangered Worker: A Labor Perspective on Timber Issues." Press package, Washington, DC, 26 May.

US Department of Agriculture and Department of Interior. 1994. *Final Supplemental Environmental Impact Statement on the Management of Habitat for Late-Successional and Old-Growth Forest Related Species within the Range of the Northern Spotted Owl.* Washington, DC: Government Printing Office.

US Department of Agriculture Office of General Counsel. 1999. Personal communication, 10 December.

US General Accounting Office. 1989. *Endangered Species: Spotted Owl Petition Beset by Problems.* GAO/RCED-89-79. Washington, DC: Government Printing Office.

US Office of Technology Assessment. 1992. *Forest Service Planning: Accommodating Uses, Producing Outputs, and Sustaining Ecosystems.* Washington, DC: Government Printing Office.

Wilkinson, Charles F., and Michael H. Anderson. 1987. *Land and Resource Planning in the National Forests.* Washington, DC: Island Press.

Wilson, Bill, Sen Wang, and David Haley. 1999. "British Columbia." In Bill Wilson et al. (eds.), *Forest Policy: International Case Studies.* 111-45. Oxon, UK: CABI Publishing.

Wilson, James Q. 1989. *Bureaucracy.* New York: Basic Books.

Wilson, Jeremy. 1998. *Talk and Log: Wilderness Politics in British Columbia.* Vancouver: UBC Press.

World Commission on Environment and Development. 1987. *Our Common Future.* Oxford: Oxford University Press.

Yaffee, Steven Lewis. 1994. *The Wisdom of the Spotted Owl.* Washington, DC: Island Press.

3
International Dynamics of North American Forest Policy: From Bilateral to Global Perspectives
Thomas R. Waggener

Canada and the United States enjoy a common and highly productive forest resource. Physically, the resource transcends the international border, gradually blending into different forest types based on geographic and environmental factors. This forest, however, is divided by sovereign political jurisdictions wherein both countries pursue multiple objectives. Each has chosen political institutions and policies governing ownership and tenure, asset transfer mechanisms, domestic manufacturing, and marketing infrastructure, including international trade, to meet its own goals.

North America represents the world's largest homogeneous market for forest products, with relatively standardized production processes and product standards. Canada, with large resources and a modest domestic marketplace, has found large and open markets in the United States. In turn, the United States has historically imported substantial amounts of forest products from Canada, supplementing domestic supplies to the economic advantage and well-being of consumers. In times of economic downturn, limited markets have led to heated but restricted disputes, such as in the recurring case of softwood lumber.

The bilateral relationship has prospered in spite of very different forest policies, including those guiding the forest sector and markets. Both nations have pursued offshore trade (exports primarily) for at least thirty years, yet the bilateral relationship in the forestry sector has remained strong and controlling. Although NAFTA was initially framed to guide this bilateral position within the context of overall national interests, forestry was little affected. Since the mid-1980s, however, this bilateral framework has come under relentless pressure from greater globalization of forestry and the expansion of trade through both multilateral and regional trade developments. Policies influencing timber supply and comparative advantage are now framed in a global context under increased competition. This new policy environment reflects both the roles of competing suppliers of timber and the rapidly changing consumer markets.

Furthermore, emerging concerns reflecting nontraditional uses of forests (nontimber and environmental values), sustainability, and the social roles of governments in community and regional welfare are leading to a fundamental reexamination of national policies in both the United States and Canada.

This chapter argues that, while the United States and Canada will continue to seek their own national interests and depend heavily on each other, national forest and trade policy formulation will increasingly require a global perspective. Policies imposed on one nation or region will likely have global implications. Efforts to balance the new dynamics of the changing globalization of forestry against the traditional bilateral relationships will be more pointed and potentially confrontational.

Canadian and US Forests in a Global Context

The forests of North America are among the most abundant and productive in the world. This forest, which existed long before the development of contemporary sovereign governments, was extensively used to meet human needs for centuries. With the advent of modern settlement, however, this forest was subjected to multiple and diverse influences, including policies intended to guide and shape resource ownership, management, and utilization. With the drawing of the international boundary between the United States and Canada along the forty-ninth parallel in 1846, this large, unbroken forest became subject to two flags and hence two different forest policy frameworks. This single act has resulted in important political, economic, and trade differences that go well beyond physical and biological resources.

The overall significance of the North American forest cannot be minimized within the global context. Forests in North America make up over 13 percent of the global forest. Forests occupy some 31 percent of the land area of the United States and almost 42 percent of that of Canada. In comparison, global forests occupy some 32 percent of the total land area. The North American share of conifer and hardwood growing stock – the forest inventory that is both factory and product for timber commodities – is 28 percent and 9 percent respectively of the global total.

By any account, these forests are important beyond the national boundaries. More than just timber, forests comprise a significant share of the natural environment and yield important social values in addition to commodities. They serve as watersheds for significant populations, shelter much of North America's flora and fauna, and include some of the most outstanding scenic, amenity, and recreational resources. Nevertheless, the forests of the United States and Canada have been the source of great economic benefits, measured in both consumption and as the engine of local and regional prosperity, employment, and income.

The Era of Bilateral American-Canadian Relations in Forestry and Forest Products

North America has historically represented the largest, most developed, homogeneous market for forest products in the world. Given the similarities in forest resources and economic/cultural conditions, the economies of the two countries exhibit a high degree of convergence in tastes when it comes to the use of wood, building styles, construction technology, and other major determinants of demand. Relatively uniform product standards have facilitated this integration of markets, including those for both solid wood products and pulp and paper products.

Compared with Canada, the United States has a significantly larger population and hence a much larger consumption of forest products. Canada, particularly British Columbia, has a relatively modest population and domestic market, but its forests and forest products industry have exceptional capacity. The United States has "needed" wood; Canada has "needed" markets. Over a long period, the two countries have become not only good neighbours but also each other's best forest products customer. The bilateral relationship has been founded almost exclusively on this simple fact.

While resources have been abundant and markets strong, the bilateral timber relationship has been a solid component of the American-Canadian partnership. Canada has been able to develop its forest resources for both the domestic market and the US market.

Table 3.1 summarizes the comparative production of timber (roundwood) in the United States and Canada for 1997. Total North American industrial timber production of conifer roundwood amounted to 40 percent of the global harvest. For conifer sawlogs and veneer logs, North America accounted for 50 percent of the global total, with 27.9 percent from the United States and 22.1 percent from Canada. The United States and Canada were less dominant for hardwood industrial sawlogs and veneer logs, harvesting 26 percent of the global hardwood sawlog and veneer log harvest.

As a percentage of global production, the US and Canadian role is strong in softwood lumber (45.9 percent), plywood (32.7 percent), particleboard (33.7 percent), wood-based pulp (52.5 percent), newsprint (44 percent), and most other grades of paper and paperboard. Due to the bilateral symmetry of the North American market, much of this forest production goes to satisfy the demands of Canadian and American consumers.

Figure 3.1 shows the percentage of US forest products originating in Canada over a three-year period. Canada is by far the dominant source for satisfying the US appetite for imported forest products, thereby enlarging US consumption far beyond the capacity of the domestic output. Canada accounted for 74 percent of the solid wood product imports in 1997.

Table 3.1

Comparative forestry statistics: US and Canada roundwood, 1997

(Thousand cubic metres)

	United States		Canada		North America		World
	Harvest	Share	Harvest	Share	Harvest	Share	Harvest
Fuelwood & charcoal	74,600	4.0%	5,319	0.3%	79,919	4%	1,854,481
Conifer fuelwood	11,625	6.4%	1,068	0.6%	12,693	7%	182,166
Nonconifer fuelwood	58,163	3.8%	4,251	0.3%	62,414	4%	1,543,702
Industrial roundwood	416,092	27.3%	185,859	12.2%	601,951	40%	1,522,758
Conifer saw & veneer logs	176,391	27.9%	139,300	22.1%	315,691	50%	631,711
Nonconifer saw & veneer logs	70,721	22.3%	12,282	3.9%	83,003	26%	316,883
Conifer pulpwood	92,870	37.7%	19,751	8.0%	112,621	46%	246,384
Nonconifer pulpwood	60,370	36.0%	11,338	6.8%	71,708	43%	167,480
Wood residues		0.0%	600	1.0%	600	1%	57,883
Other industrial roundwood	15,740	9.8%	3,188	2.0%	18,928	12%	160,299
Conifer	8,346	15.0%	937	1.7%	9,283	17%	55,645
Nonconifer	7,394	7.1%	2,251	2.2%	9,645	9%	104,654
Total roundwood	506,432	14.3%	194,366	5.5%	700,798	20%	3,537,538

Source: Food and Agriculture Organization of the United Nations. Values compiled from official FAO forestry statistics on line. http://www.fao.org

Within this group, almost all of the imports were softwood lumber. Canada also provided over 84 percent of the US imports of wood pulp products. Overall, Canada provided some 73 percent of American forest products imports in 1997.

As shown in Figure 3.2, Canada is also an important market for US forest products exports. However, compared with US imports from Canada, exports to Canada were considerably smaller. Total forest products exports in 1997 from the United States destined for Canadian markets accounted for almost 23 percent of all forest products exports. Paper and paperboard products exports to Canada were the largest product group by value, at 28 percent of all US paper and paperboard exports.

The picture is reversed from the Canadian trade perspective relative to dependence on the US market. Figure 3.3 displays the share of 1993-97 Canadian forest products exports that went to the United States.

From Figure 3.3, it is obvious that Canada is highly dependent on the United States for the large majority of wood products exports. Overall, almost 74 percent of all Canadian forest products exports was shipped to the United States in 1997. Solid wood products were the largest group exported to the United States, at over 78 percent of Canada's exports in this group. Pulp was the smallest group exported to the United States, at 47 percent of Canada's total pulp exports. Paper and paperboard exports to the United States constituted about 82 percent of Canada's exports of these products.

Perhaps more surprisingly, Canada's dependence on the United States for forest products imports is even higher than for exports. Figure 3.4 provides a summary of Canada's imports by share of total imports originating in the United States for 1993-97. The total value of Canada's forest products imports was only $5.2 billion in 1997 in contrast with exports of $28 billion.

Overall, 88.2 percent of Canada's forest products imports in 1997 originated in the United States. Over 88.5 percent of the Canadian total of solid wood products (chips, logs, poles, pilings, etc.) in this group was obtained in the United States. Some 88 percent of paper and paperboard products originated in the United States, while over 91 percent of Canada's pulp imports was obtained from the United States.

Needless to say, the bilateral relationship between Canada and the United States is extremely strong and important to both timber producers and consumers on both sides of the border. Generally, this relationship has been mutually beneficial when markets are strong and resources are relatively abundant. However, forest products markets are extremely volatile, responding to national macroeconomic shocks and periodic business cycles. Demand is greatly impacted by interest rates, which in turn impact construction – the major end use of many forest products (Waggener 1987).

Figure 3.1

Share of US forest products imports originating in Canada, 1995-7

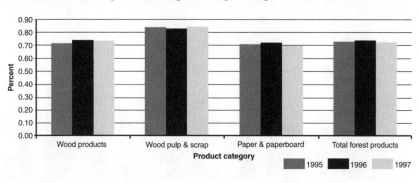

Figure 3.2

Share of US forest products exports shipped to Canada, 1995-7

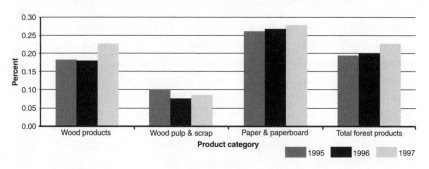

Figure 3.3

Share of Canadian forest products exports shipped to the United States, 1993-7

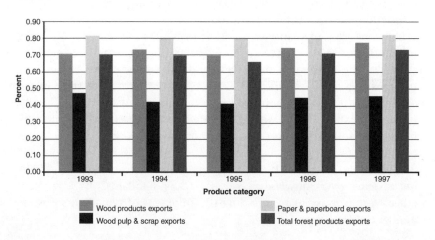

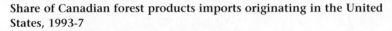

Figure 3.4

Share of Canadian forest products imports originating in the United States, 1993-7

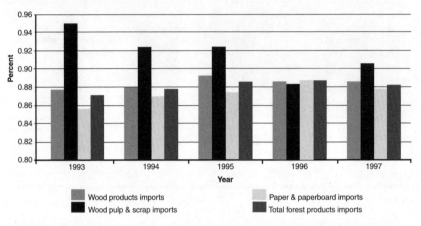

Since the early 1960s, the United States and Canada have engaged in periodic "softwood lumber wars" (Waggener 1963). As demand collapses, prices and consumption fall. Producers in both countries are placed in economically difficult conditions, profits fall, and market share shrinks. Each country is motivated to protect its own economic base and typically invokes protective measures. Most often the bilateral disputes have been over the import of Canadian softwood lumber in competition with US producers. Several appeals for tariff protection have been filed by US producers, who claim that Canadian producers have an unfair advantage because of alleged cost advantages deriving from the negotiated fee for Crown timber (Scarfe 1998; Waggener 1990). This issue has continued to be controversial (Constantino and Percy 1988; Kalt 1987). In large part, the issue can be traced to the major differences in policy related to the transfer of forest assets for utilization.

Enter the US-Canadian Free Trade Agreement (FTA) of 1989. Although not motivated to any degree by the long-standing bilateral forest products situation, the movement of Canada and the United States toward free trade was nevertheless important. Most existing "barriers" to trade in forest products were carried forward under the agreement – including the BC ban on log exports. What was potentially significant was the introduction of formal dispute resolution measures thought sufficient to lower tensions and avoid future problems such as the lumber wars (Anderson and Rugman 1990; Waggener 1990). Although ultimately successful in addressing disputes over softwood plywood grading standards, the agreement failed to resolve issues surrounding the shipment of Canadian softwood lumber to US markets.

Bilateral trade in forest products has been stressed by declining timber harvests in both the US Pacific Northwest and western Canada. In large measure, this decline is a natural consequence of the transition from liquidation of the relatively large stocks of natural old-growth timber to "second-growth," or managed, plantation-style forestry. The large volumes per acre of the natural forests cannot be replicated in intensively grown plantations that require significant investments. The reductions also reflect concerns about the past adequacy of reforestation policies in both countries (Adams et al. 1992; Reed 1990).

Other factors in reduced harvests have been broad social concern over the natural environment and the implications of taking significant portions of timberland "off base" for timber production and increasing strict (and costly) limitations on the management of remaining timberlands. It is not possible here to investigate the pros and cons of such changes, but there are important timber supply consequences. Estimates resulting from analysis of global timber supply conducted by CINTRAFOR (Perez-Garcia 1993) indicated that reductions of approximately 4 billion board feet (20 million cubic metres) were likely in both the US Pacific Northwest and western Canada.

Reduced harvests in the United States led to a successful appeal by the US softwood lumber industry for protection from Canadian softwood lumber exports in 1985. This action resulted in import duties on such shipments in 1986 (Waggener 1991). A countervailing duty was thus implemented just as the North American market was recovering from a severe recession, which had begun in 1981-82. The Canadian share of US consumption reached 30 percent in 1982-83 as the recovery slowly began. A 15 percent "surcharge" was imposed by the United States. It was subsequently modified to a Canadian export fee, then further modified as major Canadian forested provinces (led by British Columbia) modified stumpage fee (for standing timber) procedures partly to satisfy US complaints about subsidies (denied by Canadian industry) and to keep the penalty payments closer to home. The controversy abated temporarily, although over time the implicit spread between Canadian and US timber prices grew, exchange rates altered the net returns to producers, and renewed charges were again filed. The lumber war continued.

A new softwood lumber "agreement" was reached between Canada and the United States in 1996. It imposed quotas on Canadian shipments from British Columbia, Alberta, Ontario, and Quebec. Shipments over quota were subject to surcharges of $50 US per 1,000 board feet for the first 650 million board feet and $100 US for all shipments above this level. Quotas as initially allocated were estimated to range from 60 to 85 percent for the four provinces to 10 to 50 percent for lumber remanufacturers and newly established mills without prior export volumes (Miller-Freeman 1997).

This situation continued through 1997 and into 1998, still exposing the bilateral relationship to considerable stress. The US import of "fabricated structural wood" from Canada had increased from $26 million in 1996 to over $111 million in 1997 and an estimated $180 million for 1998. Canadian producers sought to "add value" to softwood lumber exports to the United States by predrilling holes for electrical wiring into lumber to be used for construction. After initially declaring such products "fabricated structural wood" instead of lumber, and therefore exempting them from the export quotas, the United States subsequently reversed the ruling (upon appeal from US producers) in April 1998. Such products are thus still classified as primary products subject to the quotas and surcharges for over-quota shipments.

This issue reflects the growing importance of international competition in forest products markets in contrast to the "typical" bilateral dispute. Both the US and Canadian industries were confronted by growing restrictions on timber harvesting, higher costs of environmental regulations, and a slumping Asian market due to recessions and financial problems throughout that region. It was no longer Canadian lumber production and US consumer markets that dictated terms of prices and consumption. The 1997 Timber Accord and policy changes in the BC Forest Code and other public policies for assisting the BC industry (e.g., the Small Business Forest Enterprise Program) caught BC producers between rising costs, declining prices, and the prospect of penalty fees under the quotas allowed into the United States duty free.

The revision of stumpage charges in British Columbia to reflect the unprofitable operations of the timber sector reduced average coastal charges by $8.10 CDN per cubic metre and by $3.50 CDN in the interior. Previously, rates averaged about $33.80 CDN on the coast and $27.87 CDN in the interior. On 27 July 1998, the United States requested arbitration of this action under the 1996 Softwood Lumber Agreement after British Columbia rejected a US proposal to reduce the approved quotas proportionately to the reductions in stumpage fees.

From Bilateral to Multilateral: Enter Mexico

Canada and the United States have certainly played a greater global role than the simple bilateral relationship described above perhaps implies. The first step in expanding the scope of formal market links was for Canada and the United States to recognize the importance of Mexico as a North American partner. Forest products trade links with many other countries continued as opportunistic markets dictated, but many are not expressly covered by formal multilateral treaties.

Building on the US-Canadian Free Trade Agreement, both countries agreed to extend free trade to Mexico, and the resulting North American

Free Trade Agreement (NAFTA), which went into effect 1 January 1994, indicated the mutual intent of establishing a broader North American market venue. Once again forestry was not a "major player" in the lead-up negotiations, although the US industry lobbied hard for a "zero-tariff" platform. US duties on most imported forest products have been either zero or very insignificant. Led by the US Forest Products Industry, the sector has lobbied in favour of reducing tariffs internationally through trade agreements and other means for all commodity and processed wood products.

Within the forestry sector, expansion of softwood lumber markets (and other products) into an emerging middle-class Mexican economy was welcomed, although there were important concerns regarding environmental consequences for Mexico's forests. Cheaper labour drawing secondary processing south across the US-Mexican border with potential re-exports into the United States and other minor issues were also expressed (Waggener 1992; Waggener 1993). One major consequence was the three-period phased reduction of Mexico's tariffs on wood products imports. US and Canadian tariffs for forest products from Mexico were already insignificant. Conifer lumber was one of the items targeted for immediate elimination of tariffs, although other products so targeted were relatively minor in terms of trade. Most major forest products commodities fell under the ten-year phase-out schedule (Schedule C), including wood in the rough (logs), some conifer lumber products, particleboard, plywood, and many joinery items.[1] Some of these items also had quotas that were carried forward during the phase-out period. Although trade in forest products with Mexico had expanded significantly in the period prior to approval of NAFTA, subsequent developments have been somewhat disappointing, as trade was reduced in the face of the economic difficulties encountered in Mexico and the reduction of demand.

Multilateral to Global: What's Up?

As noted, the forest products industry has never been strictly bilateral or even formally multilateral in the context of FTA or NAFTA. While Canada and the United States have enjoyed a special long-term love-hate relationship for forest products, each has pursued independent policies for "offshore" trade with the rest of the world. As a net importer, the United States has been considerably less dependent on exports (and more dependent on imports), while Canada essentially "lives or dies" by trade – mostly with the United States.

Table 3.2 illustrates the relative international diversification of US timber products exports (excluding pulp and paper) as of 1998. As shown, Japan is the leading export market for the US forest products industry, accounting for 27.9 percent (by value) of US exports. In total, exports to

Japan accounted for over $1.6 billion. Canada was the second leading export market, accounting for some $1.5 billion or over 26.4 percent of US export value.

Beyond Japan and Canada, US exports were fairly well distributed to Mexico, Germany, and the United Kingdom in Europe, and Hong Kong and Taiwan in Asia. No other individual country accounted for more than 7 percent of US export value.

The economic downturn of the early 1980s in the United States highlighted both the risks and the limitations of reliance on the North American domestic market. The countervailing duties imposed on Canadian softwood lumber heightened recognition in Canada that it needed to diversify its markets, mistrusting the long-term security of the US market. However, this diversification has been relatively more difficult to achieve. Table 3.3 indicates the overall structure of Canada's total forest products exports (including pulp and paper) as of 1998.

The United States accounted for $10 billion, over 84.2 percent of Canadian forest products exports, and was by far the leading export market. Asia, led by exports to Japan, accounted for almost 10 percent of Canada's export value, with Japan alone receiving some $1.1 billion or 9.7 percent of the Canadian total.

Integration of the European Community attracted significant interest in both Canada (primarily eastern Canada) and the United States (primarily the southern United States). At the same time, developments in the North Pacific (Japan, China, Taiwan, and South Korea), other ASEAN countries of

Table 3.2

US wood products exports by country, 1998

Top 10 countries	Total value	Share percent
US total ($1,000)	5,835,847	100.0
Japan	1,626,563	27.9
Canada	1,538,400	26.4
Mexico	367,575	6.3
Germany	282,201	4.8
United Kingdom	240,456	4.1
Spain	202,034	3.5
Italy	198,469	3.4
Belgium-Luxembourg	91,239	1.6
Taiwan	89,229	1.5
Hong Kong	88,496	1.5

Source: Wood Products: International Trade and Foreign Markets. 1999. United States Department of Agriculture Foreign Agriculture Service, WP-2-99.

Table 3.3

Canada wood products exports to top 10 countries, 1998

Major importers ($US million)	Total value	Share percent
Canada total	11,890	100.0
United States	10,007	84.2
Japan	1,153	9.7
United Kingdom	110	0.9
Germany	93	0.8
Italy	69	0.6
Belgium	58	0.5
Australia	56	0.5
France	45	0.4
Netherlands	38	0.3
Taiwan	37	0.3
Subtotal	11,666	98.1
All others	224	1.9

Source: Industry Canada. Values compiled from Strategis Trade Data Online. http://strategis.ic.gc.ca

Southeast Asia and the Pacific, and the growing problems with tropical hardwood supplies from Malaysia and Indonesia attracted much attention from wood products brokers. Since the early 1960s, Japan had firmly established its position as the leading global wood importer – dependent on imports for as much as two-thirds of wood consumption. The United States and Canada soon found themselves in stiff competition for the marketing of essentially homogeneous conifer products in markets from Tokyo to Taipei.

As North American forestry was facing significant adjustments in wood supply, other regions of the world were experiencing similar concerns. Sustainability was forefront in both Indonesia and Malaysia, major suppliers of tropical hardwoods. At the same time, "new" supplies were emerging in places such as Chile and New Zealand as forest plantations became very productive and competitive. Russia became an unknown as economic and political chaos replaced the somewhat predictable outflows of timber from Siberia and the Russian far east under central planning (Waggener, Schreuder, and Moffett 1992). North America suddenly realized that it was not the only global player – if it ever was. Timber had become a true global commodity in terms of both supply and demand. Each producer was in competition with all other producers; each consumer was in competition with consumers everywhere. Old assumptions and relationships were eroding, while a fundamental understanding of the "new order" was largely lacking. Nowhere was this more evident than in the historical Canadian-US bilateral relationship.

The North American role in total global forest products trade is illustrated in Table 3.4. In terms of total roundwood trade, North America accounts for about 23.4 percent of global exports (almost entirely from the United States) but only 4.2 percent of imports. For industrial roundwood, however, North America accounts for 37.6 percent of exports and 5 percent of imports. Trade in softwood lumber was more balanced, with North America accounting for almost 53 percent of exports and 37 percent of imports. As previously indicated, this is dominated by Canadian exports to the United States.

North America plays a more modest role in the global trade of wood-based panels, accounting for about 15.6 percent of exports and 14.6 percent of imports. The United States is the larger importer, while Canada provides the lion's share of panel exports. Exports are primarily softwood (conifer) panels, while imports are largely tropical hardwoods from Southeast Asia.

North America exports almost half (49.2 percent) of global wood pulp exports, with imports accounting for about 16 percent of the global total. North American exports of paper products are especially important for newsprint (52.7 percent of the global total), although this is largely Canadian exports to the United States.

Given this overall trade picture, it is evident that the United States is generally a net importer of forest products and that Canada is a large net exporter (Table 3.5). While the United States has experienced trade imbalances for solid wood products as large as $2.5 billion and trade surpluses up to $1.3 billion, Canada has seen steady growth in its total forest products trade surplus over the 1984-95 period. The US trade balance is heavily influenced by the level of revenues from softwood log exports and the deficit from imports of softwood lumber.

Globalization: Japan Case Study in Timber Markets
Economic links, of course, do not rely entirely on formal agreements or trade treaties. Where trade is relatively unrestricted, producers and consumers will find each other, and trade will commence. In the global context, Japan is dominant in terms of forest products trade, which involves both the United States and Canada as major suppliers but also many other nations, including Russia, Chile, New Zealand, and the tropical hardwood-producing nations of Southeast Asia.

Japan's importance as a global leader in forest products trade illustrates how this dynamic market has increasingly impacted the North American bilateral relationship. Since the early 1960s, Japan has experienced levels of wood consumption well beyond the domestic supply. This growing

Table 3.4

Comparative forest products statistics: United States and Canada – international trade, 1997

($US 1,000)

	United States		Canada		North America		World
	Value	Share	Value	Share	Value	Share	Value
Roundwood							
Imports	173,805	1.3%	388,794	2.9%	562,599	4.2%	13,453,040
Exports	2,208,943	21.3%	208,842	2.0%	2,417,785	23.4%	10,349,200
Industrial roundwood (conifer)							
Imports	40,016	0.8%	209,568	4.2%	249,584	5.0%	4,994,172
Exports	1,368,920	36.5%	42,910	1.1%	1,411,830	37.6%	3,752,342
Lumber							
Imports	7,555,025	26.6%	492,604	1.7%	8,047,629	28.3%	28,420,750
Exports	2,504,280	9.8%	9,393,945	36.6%	11,898,225	46.4%	25,647,700
Conifer							
Imports	7,171,940	36.1%	161,095	0.8%	7,333,035	36.9%	19,873,510
Exports	1,072,580	5.6%	9,002,910	47.2%	10,075,490	52.8%	19,083,810
Nonconifer							
Imports	383,085	4.5%	331,509	3.9%	714,594	8.4%	8,547,238
Exports	1,431,700	21.8%	391,035	6.0%	1,822,735	27.8%	6,563,886
Wood-based panels							
Imports	2,122,903	12.4%	362,353	2.1%	2,485,256	14.6%	17,063,950
Exports	1,027,783	6.1%	1,605,820	9.5%	2,633,603	15.6%	16,870,420

▲ *Table 3.4*

	United States		Canada		North America		World
	Value	Share	Value	Share	Value	Share	Value
Wood pulp							
Imports	2,670,395	15.6%	117,892	0.7%	2,788,287	16.3%	17,108,750
Exports	2,995,813	18.5%	4,969,892	30.7%	7,965,705	49.2%	16,187,700
Paper & paperboard							
Newsprint							
Imports	3,580,370	33.6%	11,630	0.1%	3,592,000	33.7%	10,646,900
Exports	552,343	5.1%	4,886,971	47.6%	5,409,314	52.7%	10,270,740
Total paper & paperboard							
Imports	11,514,180	17.2%	1,804,481	2.7%	13,318,661	19.9%	66,763,780
Exports	6,218,553	9.6%	8,856,340	13.6%	15,074,893	23.2%	65,031,620

($US 1,000)

Source: Food and Agriculture Organization of the United Nations. Values compiled from official FAO forestry statistics on line. http://www.fao.org

Table 3.5

Canada and US forest products trade balances, 1993-7

($US 1,000)

Year	United States			Canada		
	Imports	Exports	Balance	Imports	Exports	Balance
1993	16,310,381	13,537,640	-2,772,741	2,059,607	19,193,641	17,134,034
1994	18,301,482	14,411,463	-3,890,019	2,385,905	21,986,689	19,600,784
1995	22,425,589	18,148,296	-4,277,293	2,952,518	27,786,860	24,834,342
1996	23,551,699	16,641,853	-6,909,846	2,851,344	25,488,607	22,637,263
1997	24,003,249	15,698,726	-8,304,523	3,189,167	25,080,784	21,891,617

Source: FAO Yearbook Forest Products, 1999. Food and Agriculture Organization of the United Nations, Rome.

demand has been fuelled by postwar economic growth as Japan engaged in massive construction led by the need to provide replacement housing. The United States experienced significant forest damage from the Columbus Day storm of October 1962, allowing it to benefit from the Japanese need for timber as part of the salvage plan in the US Pacific Northwest (Moffett and Waggener 1992).

Japan preferred to import unprocessed logs for a number of reasons, including the desire to stabilize its own forest products sector and to adapt products to its own market requirements, which were poorly understood in North America. Only over time has the import of processed conifer lumber been successfully introduced into the Japanese market. This has involved the introduction of Western two-by-four building technology as well as greater adaptation of products to metric standards common for traditional Japanese post-and-beam construction.

The trend in Japanese imports of conifer logs is shown in Figure 3.5 for the period 1983-98. Total conifer log imports have been about 15 million cubic metres annually over this period, with a high volume of almost 18 million cubic metres achieved in 1987, when Japanese housing starts were near peak levels. Figure 3.5 also indicates the trend for the major supplier nations, highlighting the role of the United States, Canada, and the former Soviet Union. Since the late 1980s, Japan has increased the import of conifer logs from other countries, primarily New Zealand, Chile, and the "South Seas" countries.

The United States, Canada, and Russia have jointly fulfilled most of the Japanese demand for conifer logs. However, this collective share has declined from about 95 percent in the period 1983-88 to just over 82 percent for the period 1995-97. Emerging suppliers, as noted above, have been successful in capturing a growing share of this market in spite of the dominant conifer resources of North America and eastern Russia.

Figure 3.5

Japan conifer log imports by country of origin

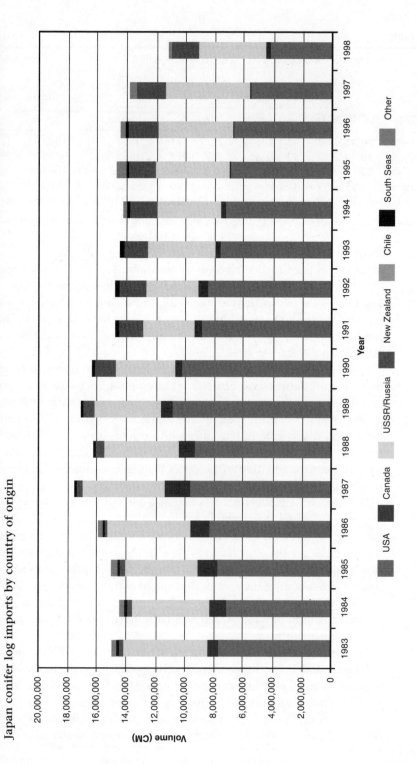

Due to log export restrictions from British Columbia, Canada's share of the Japanese conifer log market has been relatively small, reaching a peak of 10 percent in 1987 but steadily declining to almost zero by 1997. The US market share, in contrast, was consistently over 50 percent until the period 1995-97. The United States reached a peak market share of about 62 percent in 1989, when there was a strong market in Japan and Russia was on the verge of market disruptions brought on by economic and political reforms. Russia's market share fell to about 25 percent for the period 1989-92, then subsequently increased as Russia sought to export logs for immediate cash earnings. Russian market share reached about 35 percent in 1997. In that year, log exports by both the United States and Russia totalled just under 6 million cubic metres, and, for the first time since the initiation of economic reforms, the Russian total equalled that of the United States.

The trends in conifer log exports to Japan by the United States, Canada, and Russia are shown in Figure 3.6. The three dominant suppliers have seen total volumes change over time in response to both Japan's market demand cycles and changing forest resource and economic/political conditions within all four countries. Japan has sought new sources in light of supply constraints in both the United States and Russia, Russia has sought greater export earnings from timber, and the United States has seen greater restrictions placed on Pacific Northwest harvests and more restrictive export controls. Canada has chosen to withdraw from the log export market, largely preferring to promote domestic processing and the export of lumber.

As noted, over time there has been some success in expanding the export market in Japan for processed conifer lumber in partial substitution for conifer logs. The general pattern of Japan's lumber imports is given in Figure 3.7.

Figure 3.7 shows the steady growth in total conifer lumber imports by Japan from 1983 to 1997, rising from 4 million cubic metres to nearly 11 million cubic metres. Most of this increase was achieved by conifer lumber exports from Canada, which increased from 2 million cubic metres to over 6 million cubic metres for 1996 before falling slightly in 1997. The United States made modest gains in conifer lumber exports to Japan for the period 1983-89, reaching 2.6 million cubic metres. But after the peak markets of 1989, US lumber exports retreated, falling to about 1.5 million cubic metres for 1997. Russia, the other major supplier of conifer logs, played an almost insignificant role in the conifer lumber market of Japan, reflecting the state of the lumber industry in the far east region of Russia, which has limited capacity and poor technology, resulting in the lack of competitiveness. New Zealand and South Seas producers supplied small volumes of conifer lumber. However, the most significant change has been

Figure 3.6

Japan log imports from United States, Canada, and USSR/Russia (cubic metres)

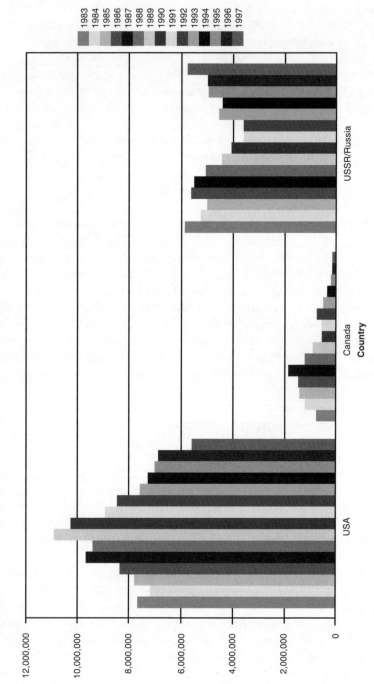

Figure 3.7

Japan conifer lumber imports by country of origin

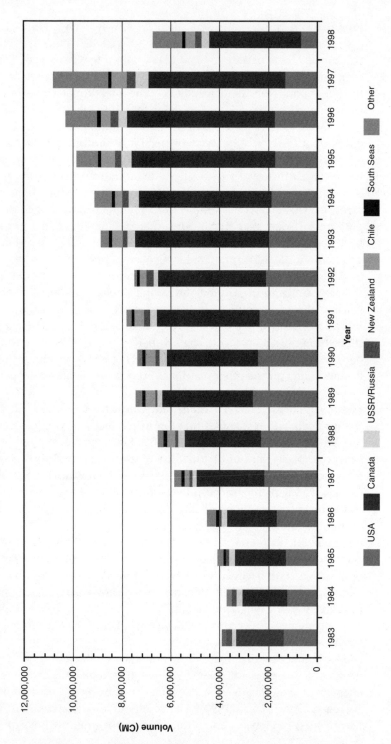

Legend: USA, Canada, USSR/Russia, New Zealand, Chile, South Seas, Other

imports from "other" suppliers since 1992, which have grown to about 2.1 million cubic metres. This amount has included small volumes from Scandinavia, a source thought unlikely only a few years ago. The Asian economic problems in 1998, however, resulted in significant declines in lumber imports from all major sources. The total dropped by over 4 billion cubic metres, to only 6.7 million cubic metres in 1998.

Trends in Japan's conifer lumber market can illustrate probable consequences for the North American bilateral relationship. Canada has experienced strong markets for lumber in the United States but has also faced increased conflict over and greater effort to restrict market access. Given this situation and the dependency of Canada on trade, perhaps it is not surprising that greater marketing efforts in Canada have shifted to gaining position in Japan, particularly for products carefully crafted to meet the needs of the Japanese market. While the United States may have been successful in the short run in reducing competition from Canadian timber in Chicago, it is finding greater competition in Tokyo! While the "big three" of Canada, the United States, and Russia have supplied about 90 percent of Japan's conifer lumber imports through 1992, the Canadian share has increased to about 60 percent, while the three-country total has slipped to under 80 percent. The US share was about 38 percent in 1986, but it has steadily declined since then, being about 16 percent in 1997. Russia's share, on the other hand, has been steady at about 4-5 percent.

Changing Forest Policy: Timber and Land-Use Changes

It is generally well recognized that forests produce many more values than just timber. Commodity production has been the leading use of North America's forests but not the only use. There has been a growing demand for noncommodity values and uses of forests in both the United States and Canada. Public policy has been reexamined and modified, with the result of significant alternations of forest land allocation between timber and nontimber management goals, together with growing restrictions on the management of forests considered as commercial timberlands. As these decisions have been implemented, timber harvests and the output of forest products have declined in both the US Pacific Northwest and western Canada. This trend is likely to continue and holds particular significance for forestry in the emerging global context.

Following major revisions of federal forest land-use policy in the United States, and the subsequent reallocations of forest lands between timber and nontimber uses (primarily for spotted owl habitat and other wildlife habitat, biodiversity, and related conservation goals), the probable consequences of these changes were estimated utilizing a global equilibrium approach (Perez-Garcia 1993). The results of this analysis illustrate the new realities of globalization and recognize the multiple links of policies implemented on a sovereign basis that extend well beyond the geographic boundaries of the initiating country.

Recent estimates by the Food and Agricultural Organization place possible global demand for timber at about 5 billion cubic metres by 2010 and 5.8 billion cubic metres by 2020. These amounts contrast with the estimated 1996 production of about 3.8 billion cubic metres. While such levels of demand are uncertain and depend on a large number of economic and technical factors, clearly total global consumption of timber products, and the derived demand for timber harvests, will certainly face upward pressures. Demands for other forest values, frequently in conflict with timber production, will also increase, possibly at significantly greater rates of change. Issues of forest sustainability, conservation of ecosystems, biodiversity, wildlife habitat, recreation, and amenity uses are increasingly conflicting and competing for forest lands. While the outcome of these pressures is far from clear, the implications can be estimated by taking into account the dynamics of global markets for timber and the responses of individual countries and producer regions initiated by changes within one country or region.

Perez-Garcia (1993) has approximated the timber supply impacts of environmental constraints on three major regions. Restrictions on land use and constraints on timber management may potentially reduce timber harvest by an average of 4 billion board feet (20 million cubic metres) annually in the US Pacific Northwest and by an equal amount in western Canada. Similar concerns about sustainability and conservation/protection of tropical hardwood forests suggest a potential reduction of an additional 4 billion board feet in Indonesia and Malaysia. This level of impact would represent about 5.5 percent of the 1992 global harvest. This would be the potential longer-term impact of present policies and regulations, exclusive of further reallocations of resources in other parts of the United States (or the world).

Using a slightly more optimistic harvest impact of about 33.3 million cubic metres (6.6 billion board feet) of softwoods in the short run, the economic equilibrium analysis[2] estimated the likely harvest (supply) impacts on a competitive global basis.

The results were estimated with and without an assumed response from Russia given the disruptions of forest policies and the unpredictable responses under prevailing economic and political instability. In a resulting equilibrium, after each country reacts to supply changes, resultant price adjustments, and their specific cost structures, it was estimated that total harvest would fall by approximately 11.4 million cubic metres, indicating that consumers worldwide would reduce consumption of forest products in the face of higher prices. Given the cost structures of individual timber-producing regions, the lowest cost suppliers would become competitive and seek to offset part of the North American reductions up to the point of the new prevailing international market prices. In the aggregate, these other competitive supply regions would increase production by about 21.9 million cubic metres. The largest responses would be (in this scenario) from Siberia (7.7 million cubic metres), Europe (4.9 million cubic metres), and the US southern forest region (2.6 million cubic metres). Higher equilibrium prices would draw additional timber from private forest ownership in the US Pacific Northwest (2.3 million cubic metres) and eastern Canada (1.2 million cubic metres). At higher prevailing prices, Japan would also increase harvest marginally (0.3 million cubic metres).

Summary
What this implies, of course, is that unilateral policies developed in the pursuit of national interest are increasingly based on external and largely unpredictable responses by other nations in the global economy. The viability of those policies is conditional on the responses and actions taken by others as well as on the "local" or domestic consequences. The United States became painfully aware of this "feedback" loop as it sought simple policy solutions to perceived problems of the growing export of unprocessed softwood logs. Since most logs were exported to Japan, there was a naive assumption that restricting the export of logs would result in a corresponding export of "value-added" lumber and lumber-based products. However, the comparative advantage in logs was not automatically transferred to lumber markets for US producers. Japan sought alternative sources of conifer logs, including supplies from Russia, New Zealand, and Scandinavia. It also expanded lumber imports, though largely from other suppliers, including Canada. US softwood lumber exports continued to decline as a share of Japanese lumber imports.

Policies imposed in one country or region now also carry likely implications for other regions and countries but may not be fully understood or considered by the initial sovereign policy makers. If a supply response from eastern Russia is foreseen,[3] for instance, it was estimated that perhaps up to five acres of Russian forests would be felled for every acre of forest "protected" in the US Pacific Northwest and western Canada. This amount is due to less substantial stocking per acre, losses in logging and transport, and less efficient recovery of useful volumes in existing facilities.

It is unlikely that this global effect on forests was well recognized or intended by US policy makers sincerely interested in spotted owl protection under US environmental laws. Thus, a well-intentioned effort in North America may in fact "export" some environmental problems and adverse forest changes. Russia, as with all countries responding to market adjustments, would implicitly intensify harvesting either in newly accessed forests or through higher cost management.

Adjustments in timber harvest are only the first direct consequence of forest land-use adjustments. A reduced harvest implies less timber processing somewhere in the global timber economy. Perez-Garcia's analysis (1993) provides an estimate of the adjustments likely to take place within the conifer lumber sector. Of the total timber reduction of 33.3 million cubic metres, some wood is allocated (competitively) to lumber, while some of it goes to other wood products and pulp and paper. The decline in lumber production was estimated at 10.8 million cubic metres. Due to price increases, total consumption would decline by approximately 3.2 million cubic metres or 29.4 percent of the decline in the North American western region. Other regions, responding to higher prices, would increase production based on their own cost structures. The greatest increases would be in Europe (3.8 million cubic metres) and the US south (2.4 million cubic metres).

What about US and Canadian Forests and Forest Policy?

Difficult choices must be made as resource scarcity increases and conflicting demands for the many goods and services expand. Can't we just manage forests for the citizens of the United States and Canada?

The above example of the potential Russian response to US conservation efforts helps us to understand the primary pursuit of national interests through unilateral forest policies (including related economic, land-use, and environmental policies). Such efforts will increasingly create feedback loops that may go unrecognized, ignored, or politically minimized under sovereign or even bilateral processes.

For example, when the United States (primarily the State of Washington) reduced log exports to Japan, it was naively assumed that Japan would simply buy the equivalent volume of sawn two-by-fours. However, a chain of market adjustments took place, with a decline in both log and lumber volumes from the United States going to Japan. Most US logs were processed for structural use in the traditional post-and-beam construction sector. North American building standards and codes for this sector left little room or rationale for greater use of two-by-fours. Japan turned to other suppliers, including Canada, Russia, and increasingly New Zealand to meet its timber needs. This potential but straightforward adjustment went largely unrecognized in debates over restrictions on US log exports.

In the past, the Canadian-US relationship was largely a North American concern, with few implications for other forest producers or timber consumers. The Asian flu of 1997-98 and the collapse of Asian markets; the transition of the former USSR; dynamic growth in China; plantations in New Zealand, Chile, and South Africa; and similar global developments demonstrate the need to weigh global links and impacts in developing sound national policy.

While Canada and the United States play out the next skirmish of the Great Softwood Lumber War, the global competitive positions of both countries are silently under continuous attack elsewhere by other players.

Efforts to protect rainforests and other environmental values in Africa and Southeast Asia, for example, will likewise frame the market realities for North America. Both production and environmental policies may be off-set by unintended and unwanted land-use and environmental impacts elsewhere as well.

How does all this affect *national* policy and forest use? Simply put, for-est management and resource conservation decisions respond largely to perceived *national* impacts and consequences. These impacts reflect both market and nonmarket values that can be provided by the limited forest resource base. Making policy decisions about national and regional prior-ities and implementing appropriate responses can no longer be viewed from a strictly national perspective. The decisions may be *national,* but the *causes* and *consequences* are increasingly global. Today's forest policy debates are increasingly about trade-offs – what are the likely advantages and disadvantages of the policy chosen? Many of those impacts, both gains and losses, fall outside national boundaries. To ignore this reality is to risk miscalculation. As Clawson (1975) aptly named his thoughtful book on forest policy, we must now ask "forests for whom and for what?" Increasingly, the answer is for the global community and for all forest values, not just commodity timber products. Is an acre of old growth pre-served in the US Pacific Northwest "worth more" than five acres of tundra in Siberia?

Forest policy will undoubtedly remain a unilateral national prerogative. A true "Global Forests Convention," as envisioned by the Rio Earth Summit in 1992, remains elusive. However, the global links and dependencies in forest management, timber production, and environmental conservation and protection are significant and important well beyond national sovereign boundaries. The means and will to effectively incorporate these impacts into sound national policy making remain difficult, since national interests and global consequences are seldom the same. This is no less true for poli-cies shaping the forests and their uses in the United States and Canada.

Notes

1 As of 1997, the phasing of tariff reductions under NAFTA progressed to where all tariffs under Schedule A were eliminated. Schedule B tariffs were reduced by 80 percent, and Schedule C tariffs were reduced by 40 percent.

2 The equilibrium analysis was computed using the CINTRAFOR Global Trade Model (CGTM), which adjusts harvest, production, and consumption in global supply and demand regions based on price and cost responses. See Perez-Garcia (1993) for a more complete discussion of this model.

3 Given the urgent demand for foreign earnings and capital, eastern Russia has increased the export of unprocessed timber even though total harvest has fallen sharply and domestic processing is only a fraction of prereform levels. Current officials perceive high-er international prices as a clear opportunity to gain additional returns from previously undeveloped forests in eastern Siberia and the Russian far east.

References

Adams, Darius, R. Alig, D. Anderson, J. Stevens, and J. Chmelik. 1992. *Future Prospects for Western Washington's Timber Supply.* Seattle, WA: College of Forest Resources, University of Washington.

Anderson, A., and A.M. Rugman. 1990. "The Dispute Settlement Mechanisms' Cases in the Canada-United States Free Trade Agreement: An Economic Evaluation." Research Programme Working Paper 34, Ontario Centre for International Business, University of Toronto.

Clawson, Marion. 1975. *Forests for Whom and for What?* Baltimore: Johns Hopkins University Press.

Constantino, L., and M. Percy. 1988. "The Political Economy of Canada-U.S. Trade in Forest Products." FEPA Working Paper 106, Forest Economics and Policy Analysis Research Unit, University of British Columbia, Vancouver.

Kalt, Joseph P. 1987. "The Political Economy of Protectionism: Tariffs and Retaliation in the Timber Industry." Discussion Paper E-87-03, Energy and Environmental Policy Center, John F. Kennedy School of Government, Harvard University, Cambridge, MA.

Miller-Freeman, Inc. 1997. "Canada-U.S. Lumber Agreement." *Widman's World Wood Review Quarterly* 4-5: 1-5.

Moffett, Jeffrey L., and T.R. Waggener. 1992. "The Development of the Japanese Wood Trade: Historical Perspective and Current Trends." CINTRAFOR Working Paper 38, Center for International Trade in Forest Products, College of Forest Resources, University of Washington, Seattle.

Perez-Garcia, John M. 1993. "Global Forestry Impacts of Reducing Softwood Supplies from North America." CINTRAFOR Working Paper 43, Center for International Trade in Forest Products, College of Forest Resources, University of Washington, Seattle.

Reed, F.L.C. 1990. "The Implications of Sustainable Development for British Columbia Forestry." Paper prepared for Price Waterhouse Third Annual British Columbia Forest Industry Conference, 27 March.

Scarfe, Brian L. 1998. "Timber Pricing and Sustainable Forestry." In Chris Tollefson (ed.), *The Wealth of Forests: Markets, Regulation, and Sustainable Forestry.* Vancouver: UBC Press.

Waggener, T.R. 1963. "An Economic Evaluation of the Softwood Lumber Industry in the Pacific Northwest." MF thesis, College of Forest Resources, University of Washington, Seattle.

—. 1987. "British Columbia and Washington State Linkages with the Pacific Rim: The Case of Forest Products and Trade." In D.E. Merrifield, R.L. Monahan, and D.K. Alper (eds.), *Growth and Cooperation in the British Columbia and Washington State Economies.* 31-74. Bellingham, WA: Western Washington University Press.

—. 1990. "Forests, Timber, and Trade: Emerging Canadian and U.S. Sector Relations under the Free Trade Agreement." Public Policy Paper 4, Canadian American Center, University of Maine, Orono.

—. 1991. "U.S.-Canada Softwood Trade: Post-1986 Developments in Changing Markets." Invited testimony, subcommittee hearings, Subcommittee on Regulation, Business, and Energy, Committee on Small Business, United States House of Representatives, Washington, DC.

—. 1992. "International Trade Policies and U.S. Trade Issues under NAFTA." Paper presented at International Policy Seminar on Forestry and Forest Products, PROAFT and World Wildlife Fund, Oaxaca, Mexico, 6 August.

—. 1993. "Forestry, Comparative Advantage, and the Status Quo: Can We Have It All under NAFTA?" Paper presented to Reunion nacional de economia forestal 93 y ii seminario nacional, "TLC y sector forestal," Colegio de Postgraduados, Montecello Edo de Mexico, 21-23 April.

Waggener, T.R., G.F. Schreuder, and J.L. Moffett. 1992. "The Pacific Rim Softwood Woodbasket: How Full Is It?" *International Trade in Forest Products Around the Pacific Rim.* 18-40. Suwon, Republic of Korea: Institute of Forestry and Forest Products, Seoul National University.

4
Firms' Responses to External Pressures for Sustainable Forest Management in British Columbia and the US Pacific Northwest

Benjamin Cashore, Ilan Vertinsky, and Rachana Raizada[1]

In the past ten years, escalating ecoforestry politics have made the world a confusing and complicated place for forest companies operating in British Columbia and the US Pacific Northwest. The domestic regulatory climate has become uncertain, international rules and institutions have placed special attention on forestry issues, and domestic and international environmental groups have targeted individual firms with boycott campaigns.

This chapter explores the ways in which three large forest companies have responded to these external demands for sustainable forest management. We review the cases of Weyerhaeuser USA operations in the US Pacific Northwest[2] and Canfor and MacMillan Bloedel (MB) operations in British Columbia. These companies all faced pressure from external interests in the past ten years, but their responses varied. All three companies initially dismissed criticism from societal and environmental groups in the late 1970s and early to mid-1980s, asserting that they already practised sustainable and responsible forestry. However, Weyerhaeuser recognized by the late 1980s that, despite its own organizational perspectives, the perceptions of societal groups were important. It readjusted its strategy, proactively setting up citizen advisory councils in an effort to take advantage of corporate social licence.[3] Once it became clear that dismissing external pressure would not stop it, Canfor at first attempted to manipulate and co-opt societal critics in the mid- to late 1980s. This strategy failed to eliminate the external pressure, and by the late 1980s the company was taking proactive measures, changing its internal environmental policies in an effort to address environmental concerns. MB took a different approach. It maintained an outwardly dismissive approach to external pressures until it reached a crisis situation in the summer of 1993, after which it explored new ways to pacify and even acquiesce to these pressures. A major change came in 1998 when the company indicated that it would alter its fundamental approach to environmental issues, announcing that it would try to be an environmental leader. Reversing years of justifying clear-cuts as ecologically appropriate in many cases, MB announced that it would end the practice in old-growth forests and that it would seek green labels for its forest products.

We explore three central themes in this chapter. First, we seek to uncover the conditions under which change in environmental responses of firms occurs. Second, we try to distinguish responses based on coercion from those based on normative value changes within a firm. Recent scholarship argues that companies undergoing value changes have a greater likelihood of leading the way with innovation and proactive measures (Vertinsky and Zietsma 1998). Third, we examine the extent to which a firm's changes in response to external pressures coincide with changes in its concept of *sustainability*. Our purpose is to contribute to an understanding of corporate responses to external criticisms of forest management.

This chapter also adds a new dimension to social scientists' exploration of the way in which ecoforestry politics have influenced public forest policy[4] by considering the roles that individual forest firms play and their strategic responses to moves by other stakeholders.[5] Firm-level analysis is important to the understanding of sustainable forestry for two reasons. First, a firm's activities are the ultimate target of these social pressures. The relationship between public policies and firm-level policy choices is crucial to understanding whether public policies have achieved the goals for which they were intended. Second, environmental groups are increasingly bypassing the state by directly targeting firms individually through negative boycott campaigns and positive voluntary certification schemes (Bernstein and Cashore 2000; Bernstein and Cashore forthcoming).

The chapter has four parts. First, it outlines a heuristic framework with which to categorize a firm's responses to external demands and its conception of sustainability. Second, it presents an overview of international forestry issues that may impact corporate choices. Third, it provides three case studies in an effort to explain corporate responses. And fourth, it assesses the policy implications of these case studies for issues of sustainable forest management and environmental protection. Research for this chapter was undertaken before Weyerhaeuser's formal purchase of MB occurred in January 2000. While it is too early to see how this acquisition will affect the former MB's operations, we speculate in the conclusion about what might transpire.

Categorizing Corporate Responses

Corporate responses to external pressures can be placed in three broad categories: fending off external demands, acquiescing to them, and proactively going beyond external demands. DiMaggio and Powell (1991: 67) identified three ways in which a firm acquiesces to external pressures through processes that they label *isomorphism*. *Coercive isomorphism* refers to changes that result when a firm is forced to change its practices. This shift is often caused by regulatory requirements or environmental groups and/or the media that effectively challenge corporate legitimacy. *Mimetic isomorphism* occurs when firms see other firms making changes that appear to address an uncertain climate. *Normative isomorphism* represents cases when a firm's internal values change as a result of external pressures. Oliver (1991) noted that firms that want to fend off external pressures may choose a variety of tactics, including compromise, avoidance, defiance, and manipulation.[6] To complete this continuum of action, a new category can be added for firms that voluntarily *overcomply*, when green values become so ingrained that the company is more advanced than societal pressures, leading the way with innovation and proaction (Vertinsky and Zietsma 1998). In the cases below, we also note that a company's responses may vary depending on the source of pressure. For example, a firm may fend off governmental pressure yet acquiesce to environmental groups.

Conceptions of Sustainability

Ever since the Brundtland Commission (World Commission on Environment and Development 1987) introduced a vague but generally appealing definition of sustainability compatible with economic growth and a liberal global trading regime,[7] the concept has been heavily promoted and studied by various groups in civil society and by scholars.[8] Within the context of forestry, sustainability is used in two ways. A limited use refers to the ability of foresters to sustain timber production, with principles derived from the German school of forestry (Johnson 1993). Under this definition, the key element is the ability to sustain a certain level of harvest in perpetuity. This concept gained strength among US foresters in the early twentieth century following growing concerns about destructive harvesting practices. British Columbia officially adopted such a definition for regulating harvests in the 1940s.

In the 1980s, however, a different conception of sustainability evolved to include not just timber production but also the biological diversity of the forest ecosystem (Noss 1989). Uncertain and incomplete information regarding forest biodiversity renders elusive a precise understanding of what should be done to achieve this version of sustainability. Differences over which outputs/values should be treated as "residuals" (Wright 1995) help to distinguish the two concepts. Those who support the former conception of sustainability tend to view biodiversity as something to be protected once harvesting goals have been taken into account; the inverse is the case for those who subscribe to the more holistic view. In this case, harvesting is treated as a residual, to be permitted only if the integrity of the forest ecosystem and biodiversity are maintained. The residual concept allows us to place responses of each company on a continuum of sustainability.

Vos (1997) highlights another important characterization of sustainability by focusing on different normative frameworks. He notes that those operating under a framework of free-market environmentalism limit sustainability solutions to market-oriented mechanisms. They believe that sustainability is best addressed through economic growth and technological innovation. Those operating under an "ecological-science" framework see the world more holistically; when economic expansion and population growth are seen as fundamental environmental problems, a broader range of policy tools is at their disposal. This distinction is important for a chapter on corporate environmentalism because it forces us to address whether or not corporate solutions are shaped by an ethos of free-market environmentalism.

International Context

The past fifteen years have witnessed increases in the related phenomena of *globalization* and *internationalization* (Bernstein and Cashore 1999). The former refers to market integration, while the latter is about the way in which international rules, institutions, and nongovernmental organizations influence domestic policy making. We briefly outline these developments and then trace their impacts on corporate decision making in the region.

Globalization has increased on a number of fronts. Market integration, corporate mergers, and the development of non-North American competitors have affected the choices that North American forest products firms have made in the past ten years. In general, firms have felt pressures to diversify beyond North America (especially to compete with low-cost producers in Brazil and other countries) and to engage in strategies of product differentiation, some of which include efforts to distinguish environmental performance from that of other companies.

At the same time, internationalization has increased dramatically. The UN Rio Earth Summit in 1992 focused attention on global forest practices. Efforts to achieve a global forest convention faltered amid north-south divisions, but a "non-legally binding authoritative statement of principles for global consensus on the management, conservation and sustainable development of all types of forests" was agreed upon (International Development Research Centre 1993). Bernstein and Cashore (2000) note that, while the Earth Summit failed to reach consensus on a set of binding international rules, it did serve as a point of germination for the concepts of biodiversity and ecosystem management, whose norms found their way into the political discourse in British Columbia and the US Pacific Northwest. Moreover, the failure to arrive thus far at an international forestry agreement has led groups to focus on voluntary certification efforts (Bernstein 1998), which provide an important context for the discussions below.

North American Certification Initiatives

Certification programs officially recognize companies and landowners who *voluntarily* operate *well-managed* or *sustainable* forest lands according to predefined criteria. The promise of market access, (possible) price premiums, and a more abstract notion of *social licence* are used to encourage companies and landowners to *certify* forest operations. Three certification programs have become dominant in the United States and Canada: the Forest Stewardship Council, backed by an environmental group; the industry-initiated Canadian Standards Association; and the American Forest and Paper Association's Sustainable Forestry Initiative.[9] We review these schemes below to better understand certification choices facing North American forest companies.

The Forest Stewardship Council (FSC)

The Forest Stewardship Council is an international certification program spearheaded by the World Wide Fund for Nature and supported by the international environmental community. Its origins can be traced to the demand of some suppliers for "green" wood as well as increased disillusionment with the stalled global forest convention negotiations (Bernstein and Cashore forthcoming).

The FSC has ten "principles and criteria" that are performance based and broad in scope, covering tenure and use rights (principle 2), community relations and workers' rights (3), environmental impact (6), management plans (7), monitoring (8), and the maintenance of high conservation values (9). A tenth principle was initiated to cover plantations.[10]

Governance is structured at FSC by a *three-chamber* format in which economic, social, and environmental interests have their own chambers. Each chamber contains 50 percent representation from northern and southern countries (Bruce 1998: Chapter 2). All three chambers must agree on policy choices.

Standards to implement the ten principles are carried out at the regional level, where processes have been developed in the United States and Canada, including the US south, the US Pacific region, British Columbia, and the Maritimes. Owing to new membership and efforts to achieve consensus, many of these regional bodies are years away from achieving regional standards.[11] FSC accredits certifiers rather than doing the certification itself. In the United States, two organizations are accredited to certify for the FSC: the Rainforest Alliance Smart Wood Program and Scientific Certification Systems Forest Conservation Program (Forest Stewardship Council 1999a). These certifiers in turn license affiliates in different US states to carry out on-the-ground accreditation. In the absence of regional standards, affiliates develop provisional standards that are approved by the international FSC body, located in Oaxaca, Mexico.

The FSC is considered a *third-party* system because it is independent of the companies and landowners seeking certification.

As FSC has gained attention in recent years, its membership dynamics have likewise changed. For example, many BC industry organizations, including the BC Forest Alliance and the Industrial Wood and Allied Workers union, have joined FSC in an effort to influence organizational choices and to gain societal credibility. Reflecting these changes, FSC has altered principle 9 from not allowing logging in old-growth forests to maintaining high conservation values. Most domestic and transnational environmental groups support the FSC.[12]

The US American Forest and Paper Association Sustainable Forestry Initiative (SFI)

The US Sustainable Forestry Initiative is a program of the American Forest and Paper Association (AF&PA), whose 200 members own 90 percent of industrial forest land in the United States (Hansen and Juslin 1999). The origins of SFI are actually earlier than those of FSC and can be traced to an interest in addressing the polarized atmosphere that had developed in the United States over forestry issues in the late 1980s and early 1990s. SFI was formally created after an AF&PA-commissioned study revealed different perceptions on the part of civil society and industry about whether sustainable forestry was being practised (Wallinger 1995).[13]

SFI is focused less on performance than on procedures. It uses implementation guidelines, objectives, and performance measures (Hansen and Juslin 1999: 19). Performance requirements are largely limited to following riparian best management practices (BMPs), legal requirements, and regeneration requirements.[14] Procedurally, member companies are required to file a report with SFI regarding their forest management plans and the objectives that they are addressing. Specific company data are not reported. Instead, information is aggregated and given to a panel of experts for review. The emphasis is on continual improvement. SFI has been used in efforts to improve the industry profile, including advertising campaigns. At the same time, SFI is adapting its program to meet broader societal scrutiny. The independent panel has gained increased autonomy to make recommendations, and voluntary third-party audits are now permitted.[15] SFI, as an associational program that accredits its own members, is considered a *second-party* program.

The Canadian Standards Association (CSA) Forest Program

The origins of the Canadian Standards Association Forest Program can be traced to the interest by the Canadian Pulp and Paper Association (CPPA) to develop a program recognizing existing sustainable forestry practices in Canada, which were coming under increased attack in Europe by those who wanted to stop logging in Canadian old-growth forests (Stanbury, Vertinsky, and Wilson 1995). Much of the European effort was focused on BC forests, in which most of the forest type slated for logging is characterized as old growth. To justify continued harvesting of old growth, the industry needed a mechanism to validate the sustainability of this and other practices.

The focus is largely, but not exclusively, on process. Instead of developing a program under the CPPA or another industry body, industry turned to the reputed CSA to develop "a systems based approach to sustainable forest management" (Hansen and Juslin 1999: 20) that mirrors in many ways the procedures of the International Standards Organization. This approach is based on individual companies establishing internal "environmental management systems" (EMS) (Moffat 1998: 39). Two different implementation and auditing standards have been developed under CSA. The first "describes the design and implementation of a forest management system that includes environmental, economic, and social and cultural aspects," while the second details auditing requirements (Hansen and Juslin 1999: 20). CSA-approved companies are to turn to criteria and indicators developed by the Canadian Council of Forest Ministers (CCFM) when developing their EMS.

The CSA program varies somewhat from SFI in that it has broad requirements for consultation (Table 4.1), which some industry officials say are more onerous than FSC standards (personal interviews).[16] In fact, there is an on-the-ground performance element to CSA in that a company must "address the six criteria described in the CCFM Criteria and Indicators for sustainable forest management in Canada" (Moffat 1998: 40). However, companies have significant latitude in how their EMS address these criteria; companies are also given the authority to develop their own criteria and indicators, as long as they fit within the CCFM framework (Moffat 1998: 40).

Overall, the CSA emphasis is on firm-level processes and continual improvement. The program is best placed under the *second-party* category, even though technically CSA is outside any industry associational system. In practice, it operates as an industry organization. The CSA rules were developed by industry organizations; environmental groups withdrew from the process early in their development. The CSA program fits within the conception that industry should regulate itself with considerable transparency.

In Canada, some companies have chosen to become certified under both CSA and FSC, but it is clear that each program represents a different notion of environmental governance. Indeed, efforts to merge these programs in the mid-1990s failed, in large part because the two conceptions of private sector governance could not be reconciled (Gale and Burda 1997). Canadian companies may choose both in the short term to gain social licence, but in the long run these schemes can be seen as competing

Table 4.1

Comparison of FSC, SFI, and CSA forest certification schemes

Program	FSC	SFI	CSA
Scope	International	National	National
Origination	Environmental groups	Industry	Industry
Performance- or systems-based	Performance	System	System
Verification	Third party	Second party, third voluntary	Second party
Public involvement in programs and policies	Yes	No	Limited
Public involvement in company accreditation	Yes	No	Yes
Chain of custody	Yes	No	No
Ecolabel	Yes	No	No

Source: Adapted from Moffat (1998: 152); from a compilation of certification schemes by Mark Rickenbach, the Sustainable Forestry Partnership at Oregon State University; and from a presentation by Rick Fletcher and Michael Washburn to the Society of American Foresters Annual Meeting in Portland, 13 September 1999.

with one another for legitimacy. In the United States, SFI and FSC are more clearly competitors. No member of the AF&PA has announced its intention to become FSC certified, though some are waiting to see which program emerges. We trace each company's response to certification in the discussions below.

The Cases

Weyerhaeuser USA

Weyerhaeuser is a North American forest products firm that manages 5.3 million acres (2.1 million hectares) of its own forest land in the United States and an additional 22.9 million acres (9.3 million hectares) of public forest land in Canada through long-term tenure arrangements (Weyerhaeuser USA 1997: 27). Its divisions include Timberlands, with 2,600 employees; Wood Products, with 14,200 employees; Pulp, Paper, and Packaging and Recycling with 18,200 employees; and real estate and mortgage divisions (Weyerhaeuser USA 1997: 28).

Unlike most forest companies that "cut and ran" after harvesting in the early 1900s, Weyerhaeuser held on to its forest lands, believing that active management might lead to profitable forestry. Such a decision ran counter to the prevailing wisdom of the time that tax policy and slow natural regeneration made second-growth management uneconomical (Day, Hart, and Milstein 1998). The company began this experiment first by seeding and then by planting and intensive forest management. Building on these practices, a model of high-yield forestry (HYF) began to be practised on Weyerhaeuser lands in the 1960s (Day, Hart, and Milstein 1998). At this point, Weyerhaeuser's approach to harvesting fit within the first concept of sustainability outlined above: practising sustained timber yield that relied on science for management and intensive forestry techniques. At the same time, the decision to keep its forest land gave Weyerhaeuser the capacity to respond relatively quickly to environmental issues, first in the 1960s and again in a comprehensive way in the late 1980s and 1990s.

It was during the 1960s, 1970s, and 1980s that environmental forestry concerns in the United States were addressed through an array of legislative initiatives (Cashore 1997, 1999). Most of these statutory requirements and rules were geared toward public forest lands, where environmental-industry conflicts were felt most strongly. Many companies with private forest lands felt relatively shielded from this type of pressure and tended to ignore these developments (Day, Hart, and Milstein 1998: 16-6). Rules governing private forestry at the state level were written to allow managerial flexibility, with few opportunities for litigation (Cashore 1999; Salazar 1989). Even federal statutes such as the Endangered Species Act (ESA) applied quite differently on public lands as opposed to private lands. Public land managers are required to provide a program for species recovery, while private landowners' actions must not result in the "taking" of an endangered species. Moreover, private owners of forest lands have the flexibility to develop a "habitat conservation plan" (HCP) that allows for "incidental" takings under ESA.[17]

Weyerhaeuser's Initial Response

Despite less attention to private forest lands, Weyerhaeuser did not dismiss or ignore these societal pressures in the 1960s. Instead, key Weyerhaeuser officials moved to strategically accommodate these pressures, recognizing explicitly that addressing environmental concerns could improve not only the ecological impact of forest operations but also its corporate image. Indeed, Weyerhaeuser was the first US forest products company to develop an environmental policy in 1971 (Weyerhaeuser USA 1995: 7).

By the late 1980s, public attention had expanded its focus to private forest land, with an increasing awareness of the effects of private forestry on fish and wildlife populations and water quality. The public and environmental groups were also concerned about the effects of the much less strict regulations governing private forest land[18] compared with those affecting public forests. Despite its stance in the 1960s, Weyerhaeuser soon came under public and environmental group scrutiny and criticism.

Its initial response was to dismiss such pressures by arguing that it already practised responsible and sustainable forestry, pointing to its 1971 environmental policy (Weyerhaeuser USA 1995: 7). Weyerhaeuser emphasized its forest regeneration programs and used its scientific data to deflect charges that it was not an environmental forest steward. However, these initial reactions failed to convince the public and environmental groups.

In 1989, Weyerhaeuser hired a new CEO, John Creighton Jr., whose job was to turn around poor financial performance. Creighton linked financial performance to regaining public trust, which he believed the company had lost. He embarked on a number of environmental initiatives. In 1991, he established an Environmental Council comprised of key Weyerhaeuser officials to facilitate the development of an environmental policy (Weyerhaeuser USA 1995). In the US Pacific Northwest, Creighton initiated a series of preliminary meetings with interested stakeholders in 1994 (Weyerhaeuser USA 1995: 23). The new CEO soon learned that the company was not acting as a steward of the resource "in the eyes of the people" (Day, Hart, and Milstein 1998: 16-6) and that the public viewed both public and private forests as providing key social and environmental values. As Day, Hart, and Milstein note (1998: 16-6), "Weyerhaeuser's attempts to justify its practices through mounds of data had failed to address fundamental public fears over the health and future of forests and ecosystems." Weyerhaeuser's *1994 Annual Environmental Performance Report* stated that "We learned that the public wants us to look at the whole forest, including non-timber resources. People want us to leave buffers around streams and more standing trees for wildlife habitat" (Weyerhaeuser USA 1995).

Weyerhaeuser Forestry and Sustainability

The result, as Day, Hart, and Milstein (1998) describe, was that Weyerhaeuser's conception of sustainable forestry moved further away from the first conception to one in which a variety of resources was accommodated. HYF evolved into a process called "Weyerhaeuser Forestry." Weyerhaeuser responded to external pressures with this new conception of sustainability, incorporating them into operational decision processes through the creation of regional forest councils as well as through the use of town hall meetings in Weyerhaeuser communities.

A consequence of this new approach was the development of a "stewardship statement" and a number of strategies aimed to guide the company in implementing its "more holistic model of sustainable forestry" (Day, Hart, and Milstein 1998). The stewardship statements include commitments to water quality, fish and wildlife habitat, soil productivity, biodiversity, and aesthetic, cultural, and historical values. The Weyerhaeuser Foundation also oriented many of its grants toward improving environmental stewardship and conservation projects (Weyerhaeuser USA 1995: 24). It proactively sought out partnerships with environmental and other non-governmental interests, which included its participation in the creation of new wildlife refuges.

In the mid- to late 1990s, Weyerhaeuser took on a more proactive role in the protection of endangered species through the development of habitat conservation plans and watershed analyses (Day, Hart, and Milstein 1998). The flexibility provided by the ESA for species protection allowed Weyerhaeuser to become proactive, as well as to minimize the economic consequences of a strict application of the law (Cashore and Vertinsky 2000; Day, Hart, and Milstein 1998).

Weyerhaeuser Forestry was put to the test in 1989 when the spotted owl was listed as a threatened species in the US Pacific Northwest. The management plan proposed by the US Fish and Wildlife Service would have defined a no-cut zone 1.2 miles in radius around each identified spotted owl nest. Through its HCP approach, however, Weyerhaeuser gained federal approval for a spotted owl conservation plan that allowed more flexible use of its forest lands, opening up timber assets that had previously been defined as off-limits under the "owl circle" management plan.

Weyerhaeuser also actively supported the American Forest and Paper Association's Sustainable Forestry Initiative (SFI) (Weyerhaeuser USA 1997: 10). In accordance with SFI principles, Weyerhaeuser initiated internal audits. As of late 1999, Weyerhaeuser had not ruled out achieving certification under other programs, such as the Forest Stewardship Council. Weyerhaeuser officials were actively monitoring the development of and interest in different certification programs to see which kinds of choices they might have to make in the future. Weyerhaeuser officials noted in the summer of 1999 that they believed the costs of implementing SFI and FSC programs to be roughly similar, indicating that expenses for certification would not by themselves be a determining factor in future policy choices (personal interviews).

On the pulp mill side of operations, CEO Creighton announced plans in 1995 that would see the company overcomply with EPA pollution controls ("New Growth" 1995). It similarly took advantage of an EPA program to allow regulatory flexibility when a company performs above regulatory requirements ("Weyerhaeuser Signs Project XL Agreement" 1997). This did not mean that Weyerhaeuser was able to eliminate all instances of non-compliance, but it did succeed in reducing these rates (Lambert 1996). Indeed, the company noted in 1995 that it "must improve its environmental compliance performance" (Weyerhaeuser USA 1995: 22).

By the end of the 1990s, Weyerhaeuser had clearly embarked on a new approach to forestry management and public input. As a profit-maximizing firm, Weyerhaeuser still held timber production paramount, but the company now tried to achieve goals to "protect, maintain, or enhance other important environmental values" (Weyerhaeuser Timberlands USA 1997).

This change in approach did not mean that Weyerhaeuser was immune to globalization. Illustrative of increasing capital mobility associated with globalization, Weyerhaeuser expanded its forest operations to a wide array of forest types and legal/regulatory settings. Weyerhaeuser has sold or closed some of its US Pacific Northwest operations (Batsell 1998; "Roseburg" 1995; "Weyerhaeuser to Cut Production" 1998) while increasing assets in the US south ("Weyerhaeuser Builds New Plant" 1995), where most states do not have forest practices acts, and for the first time moving outside North America to New Zealand (Brown 1997; Hall 1997; "Concern" 1995), where plantation forestry is dominant.

Making generalizations about a large, vertically integrated forest company is fraught with difficulties. Nonetheless, the story of Weyerhaeuser's approach to external interests in the US Pacific Northwest (especially since the late 1980s) has been to take an innovative and proactive approach, although within the constraints of profit maximization in an increasingly globalized and highly competitive industry. Unlike the experience of MacMillan Bloedel below, Weyerhaeuser was quick to see the potential competitive advantage of being green, an area that it continues to explore (Day, Hart, and Milstein 1998: 16-12).

Canfor

Canfor is a BC-based forest products company that specializes in pulp and paper and wood products. It owns 100 percent of Canadian Forest Products Limited (which produces pulp and kraft paper, lumber, hardboard, and fibre) and 50 percent of Howe Sound Pulp and Paper Limited (pulp and newsprint) (Canfor Corporation 1998: 5). It employs 4,300 persons in its forest products operations and about 1,300 more in its affiliates. It harvests timber from publicly owned forest lands through timber licence agreements, which give the company forest management responsibilities in exchange for a secure supply of fibre. When the company was first the target of external demands to improve its sustainable forestry management in 1985, forest companies in the province operated under a non-legal regulatory regime in which agency officials enjoyed a great deal of discretion in interpreting, administering, and enforcing regulations (Cashore 1997; Wilson 1990).[19] In 1985, Canfor felt external pressure from "civil society" stakeholders, as organized environmental interests and citizen groups targeted Canfor for logging in old-growth forests and for its pulp mills' air pollution (Raizada 1998: 163).

Canfor's Response to External Pressure

The company's initial response was to avoid and defy these social pressures. Canfor avoided the issues by arguing that old-growth protection would have serious negative consequences on the provincial forest economy. As well, like Weyerhaeuser, Canfor sought to dismiss and pacify criticism by asserting that its harvesting methods were environmentally sustainable.[20] Meanwhile, Canfor used its close relationship with provincial officials to avoid meeting pollution-level targets by proposing management plans that allowed high sulfur pollution levels to continue (Raizada 1998: 164). The nonlegal/discretionary regulatory style characterizing BC forest policy posed little threat that the courts would require Canfor to meet these targets (Hoberg 1993).

In 1987, Canfor's Howe Sound pulp mill's air pollution came under increased attention from media outlets, organized environmental groups, and local citizens, who jointly argued that Canfor should at least meet its permit requirements (Raizada 1998: 165).[21] Despite the increased attention, Canfor attempted to defy external pressures by adding manipulation strategies to its previous defiance and avoidance techniques, holding a series of public meetings in order to resist these pressures. This time, however, the company began to lose public credibility. Government officials were compelled to respond, and environmental groups threatened to launch an international boycott campaign, which raised the possibility that the firm's profits would be affected if economic stakeholders became involved (Raizada 1998).

With the threat of increased public pressure and the entry of economic stakeholders who might affect profits, Canfor began to alter its approach, taking proactive measures. The company stopped defending its level of pulp mill pollution and announced that it would build a state-of-the-art, environmentally friendly pulp mill. Here the firm turned to the promise of technological innovation to address its environmental problems, a solution designed to increase profits and improve environmental quality. Canfor was able to move quickly owing to a cooperative team-oriented organizational structure that facilitated the relatively quick implementation of senior management decisions (Raizada 1998). CEO Peter Bentley promised that the project would enable Canfor to deal completely with its environmental problems, and the vice-chairman acknowledged the role of external pressure in forcing this change: "some of the publicity" about Canfor's Howe Sound pulp mill "has not been entirely positive. It is very tempting to blame certain groups or the media for this kind of publicity, but instead we have been carrying out our improvement program there and making information about progress available to the public and the employees as we go along" (Canfor Corporation 1988: 5). These firm-level changes in strategic response appeared to reduce the threat to changes in the discretionary regulatory regime. However, the complicated world of ecoforestry politics would show that Canfor's experience with external pressure had just begun. Just as Canfor "successfully" addressed air pollution concerns, its Howe Sound pulp mill's water pollution became a provincial, national, and international issue.[22] Canfor was singled out by environmental groups in November 1988 when dioxins discovered in the shellfish near the mill led to the closure of the shellfish industry in the area. Canfor's close relationship with provincial regulatory agencies again showed signs of stress, and a series of new regulations and fines governing pulp mill pollution was announced. In May 1989, the provincial Ministry of the Environment (MOE) announced new regulations to control the discharge of organochlorines. Regulations governing the levels of BOD (biochemical oxygen demand) and TSS (total suspended solids) in pulp mill effluent were tightened, requiring all mills to have a system of secondary treatment. Moreover, MOE began to flex its regulatory muscle, increasing fines for violations of the Waste Management Act. Revisions were made to the Federal Pulp and Paper Effluent Regulations as new international boycott campaigns were plotted.

Canfor responded to these pressures in April 1989 with innovative measures, creating the position of vice-president in charge of environment and energy and introducing its new environmental policy. The Department of Environment and Energy was to perform environmental audits of all the company's operations to determine how well the operations complied with regulatory requirements and corporate standards and policies (Raizada 1998: 178). Canfor also set a goal to reduce wood waste at its forestry operations by 50 percent. The company promoted its new positions and environmental policy, arguing that it had again made changes in response to public concerns. It also sought to accommodate Greenpeace through meetings and seminars (Raizada 1998: 179).

This transition from dismissal and pacification to manipulation and then innovation and proaction happened as a complete change in the regulatory regime appeared to be imminent. Canfor was spurred on by the fear that environmental groups might succeed in forcing purchasers of its products to look elsewhere and by the desire to maintain the existing nondiscretionary regulatory style. This strategy seemed to be successful in the short run, for the regulatory style remained intact.

The Regulatory Environment Changes
Canfor, however, faced yet more uncertainty and possible regulatory change with the election of the social democratic New Democratic Party (NDP) in 1991. The NDP ushered in an array of new consensus-oriented participatory institutions in which, in many forestry issues, forestry organizations were now but one player among a wide range of societal and economic stakeholders (Cashore 1997). The BC forest industry was increasingly forced

to share influence in the policy-making process with environmental groups and other social actors (Cashore 1997; Wilson 1998). The regulatory style became increasingly complex, but consider-able discretion remained, although, as Hoberg notes in Chapter 2 in this volume, there remained only limited opportunities for environmental groups to litigate.

The new government quickly moved to revise pulp mill pollution regulations. Environment minister John Cashore announced that the government would introduce strict regulations to reduce pulp mill effluent.[23] The federal government also considered its own pulp mill regulations. Canfor seized on this overlapping of provincial and federal jurisdictions to argue that British Columbia should cede authority to the federal government, which was considering less stringent regulations. At this juncture, Canfor sought to compromise with BC government officials but appeared to move back from its proactive response. It did not argue against environmental regulations per se, but it pointed out that the proposed provincial AOX (adsorbable organic halogen) regulations would not improve the environment: "The provincial requirement to completely eliminate AOX discharge in mill effluent by December 31, 2002 ... [will cause] substantial expenditures which cannot be justified on environmental grounds" (Canfor Corporation 1992: 37).

Canfor was now playing the role of compromiser, hoping to minimize pressure from government officials – far from the proactive stance that it took previously. This change from being proactive under a minimal and discretionary regulatory regime to being compromising under a more strict and complex regime raises questions about the conditions under which firms are likely to be proactive.

However, as Canfor sought to compromise with the BC government on its pulp mill regulations, a new European environmental campaign began to increase awareness of the presence of dioxins and furans in Canadian pulp and paper products. Greenpeace Germany garnered support from leading German publishers to demand alternatives to chlorine-bleached pulp ("Greenpeace Turns up Pressure on Germany" 1993). Now Canfor faced a situation in which government, social interests, and some economic actors applied pressure for not only Canfor but also the entire BC industry to change. As world opposition to chlorine-bleached pulp grew, international attention expanded to include forest practices and old-growth preservation (Cashore 1997).[24] Canfor faced pressure on a number of fronts: pulp mill pollution, the new government's promise of a Forest Practices Code, and old-growth wilderness protection issues. Canfor did not fight either the code or increased wilderness preservation (contrary to its initial mid-1980s stance). Instead, the firm bargained by advocating the type of code that it preferred and participated in land-use processes with other organized interests, including environmental groups.

Canfor Becomes Proactive

Owing to the wide range of scrutiny from a multitude of external actors, Canfor made a series of decisions ranging from green value changes to proaction and innovation. For example, the company announced that it would "lead the way" in investing in alternative bleaching processes for the express purpose of meeting "the rapidly expanding demand in Europe for pulps bleached without the use of elemental chlorine or any chlorine compounds."[25] Canfor took a similar approach to sustainable logging practices, which now dominated domestic and international attention (and which focused largely on MacMillan Bloedel). To address and pacify the concerns of its customers in Europe, Canfor held information sessions and explained changes that it had made (Raizada 1998: 191). In response to the impending BC Forest Practices Code, Canfor developed its own "Forest Practices Compliance Policy" (Raizada 1998: 205). It now operates a "Forest Practices Performance Review Program" in which audits are undertaken for "compliance with government legislation, Canfor's Forest Stewardship Policy, forest industry standards and generally accepted good forestry practices" (Canfor Corporation 1998: 9).

Similarly, Canfor took a strong interest in sustainable forestry certification efforts. As with Weyerhaeuser, Canfor's concept of sustainable forestry now moved toward the more holistic definition, although harvesting remained the dominant concern. Indeed, its efforts to use the Forest Practices Code as a marketing tool and its move toward green labelling made its policy options consistent with the ethos of free-market environmentalism.

Canfor's responses also illustrate how a firm may respond to external pressures on the same issue at the same time in different ways. While Canfor made innovative responses to environmental group and economic pressures, it compromised with the BC government by showing regulators that firms can proactively address environmental protection issues without increasing the regulatory burden.

Canfor was also relatively quick to move its operations toward certification under the CSA Sustainable Forest Management Standard. The firm noted in early 1998 that it planned to apply for certification in the Prince George Timber Supply Area "as soon as the CSA registration process is fully in place and Canfor has completed the required public involvement process" (Canfor Corporation 1998: 9). The company sought to adopt initiatives that could address profit-maximizing objectives and increase its legitimacy. Certification was such an initiative. As the company noted, "Certification will ... help address market concerns expressed ... by pulp and paper customers in Europe who are under pressure to ensure their suppliers are managing their forests responsibly. Within Canfor's broad, domestic community, certification should also further strengthen public confidence in Canfor's forest stewardship" (Canfor Corporation 1998: 9).

Supporting private forest certification helped Canfor to maintain its image as an environmentally proactive company and to focus the attention of its critics on private firm-level initiatives rather than on the more unstable and uncertain public policy regulations. In 1998, for example, Canfor successfully lobbied to reduce the Forest Practices Code procedural rules and stumpage fees (Hunter and Hogben 1998).

MacMillan Bloedel (MB)

MacMillan Bloedel is Canada's largest forest products company (in terms of sales and market capitalization), with integrated forest operations in the United States, Canada, and Mexico. The company also has marketing facilities in North America and Japan and a number of joint ventures in North America. MB has three core business segments: building materials, paper, and packaging. It manages 2 million hectares of productive timberlands, half of which are located in British Columbia. Most of this land is in area-based timber farm licences, but it also has significant holdings of private forest land in the province. Thirty-eight percent of MB's forest land is second growth, a high percentage compared with most BC forest companies. This gives MB relatively more flexibility in addressing old-growth issues.[26] It has recently acquired a plant in Mexico and has a long record of direct foreign investment in Brazil, Australia, Holland, and the United Kingdom.

MB's hierarchical organizational structure encourages employee competition and thus discourages somewhat the cooperative team-oriented approach that Canfor and Weyerhaeuser established. Much of MB's organizational structure can be traced to the autocratic nature of H.R. MacMillan (see Drushka 1995). MacMillan instituted a rule that family generational links were explicitly forbidden, a rule that remains in place today. He also institutionalized a highly competitive system of rewards and benefits that appeared to discourage a cooperative team approach to problem resolution. MB had a proportionately higher number of top managers, with over thirty executives holding the title of vice-president, senior vice-president, or executive vice-president in 1995 (as opposed to fifteen for Canfor). Partly as a consequence of its bureaucratic structure, MB historically tended to take a legalistic approach to its operations (Zimmerman 1997: 65).

Like Weyerhaeuser, MB embraced the timber-oriented definition of sustainability relatively early. In 1938, MB became the first BC company to reforest using scientifically based timber management practices. In 1961, MB started a more comprehensive forest management program designed to ensure the security of its future wood supply. This program evolved in 1979 into the Designed Forest System: a series of silvicultural treatments applied to individual stands of trees at various stages of development to optimize their growth. Attention to improved silviculture/sustainability is constant, and in 1985 new methods were developed for comparing the value of different silvicultural treatments to ensure that the most beneficial ones were given priority.

MB and External Pressures for Change

MB's experience with environmental group pressure dates back to the 1970s. During this time and into the 1980s, the concern was primarily with preservation of old-growth forests. One of the first areas of conflict to receive public attention was a dispute between environmentalists and Aboriginal bands over MB's plans to log old-growth forests on Meares Island in Clayoquot Sound and South Moresby in the Queen Charlotte Islands.[27] MB responded by attempting to avoid these pressures. It withdrew from a Meares Island planning team in 1983, arguing that the company's goals were being ignored. MB then relied on its close relationship with the Social Credit government, which subsequently ignored the planning team's recommendations for protection. However, environmental and Aboriginal groups sustained pressure through civil disobedience and through the courts. In March 1985, the BC Provincial Court of Appeals enjoined MB from logging Meares Island until Aboriginal land claims were resolved.

In the fall of 1985, MB logging operations on South Moresby came under sustained protest as well from environmental groups and Aboriginal groups. Despite this scrutiny, MB did little to address the environmental concerns regarding its forestry practices. Its annual report focused on economically viable timber, asserting that planned reforestation backed by scientific research would do an adequate job of renewing the forests. MB's concept of sustainability remained one of timber production, which influenced the way in which the company addressed these pressures. For example, MB warned that protection of areas such as Meares Island and South Moresby would jeopardize thousands of jobs.

At this juncture, MB maintained that its forest practices were sound but acknowledged that it was doing a poor job of communicating its position. In 1987, its annual review stated that "MB is very aware that a balance must be sought between its ability to compete effectively with the world's largest forest corporations and its responsibilities to its many stakeholders-shareholders, employees, customers, suppliers, governments, and members of the public ... It is not enough that the Company believes itself to be acting in the best interests of its stakeholders – that must also be the perception" (MacMillan Bloedel Ltd. 1987: 4, 8).

Since MB's actions at this point were focused on perception rather than substance, the company developed a public involvement strategy directed mostly toward an intensive communications and advertising program to tell the people of British Columbia that it was "a responsible guardian of the forests it manages" (Raizada 1998). It maintained an approach that was outwardly dismissive of environmental group pressures. In August 1988, for example, MB's assistant chief forester stated that British Columbia's environmentalists had "a simple and unfinishable agenda that would see all resource industries brought to a standstill ... Believe me, there's no end to it" (personal interview). In 1990, CEO Ray Smith explained the implications for MB of this multiplicity of interest groups: "The diversity of single-issue causes virtually ensures there can be no solution ... This group is worried about the aesthetic values of the forest. That group is worried about the microorganism substrata in the soil ... There's an almost limitless meridian of different single causes, most of which don't agree with each other, so to try to bring all this together and to try to find a middle ground is very difficult, if not impossible" (personal interview).

MB maintained its dismissive approach through the late 1980s as scrutiny over old-growth issues, forest management practices, and pulp mill pollution increased domestically and internationally. Indeed, national attention increased so dramatically on South Moresby that a deal was brokered in which a national park would be created and MB would receive compensation.

By the late 1980s, multiple conflicts were breaking out. The use of chlorine in pulp production was gaining international attention, largely through the efforts of Greenpeace.

In November 1988, the federal government closed three shellfish-harvesting areas adjacent to coastal pulp mills in British Columbia due to high dioxin levels, thus bringing more public attention to the pollution problems created by MB's pulp mills. In May 1989, the provincial MOE announced a schedule of controls on chlorinated organic compounds discharged from pulp mills. In July of that year, MB was informed of the plan to establish mill-specific regulations for its Alberni mills.

MB's reaction to the new regulations was to fight them aggressively, arguing that the company would probably be forced to vacate facilities. But this time the government was not siding with MB: the BC environment minister said that, if MB was hinting that it wanted special treatment to save jobs, then the company should forget it. "I think there may be some pressure put on me by pulp mill owners to relax certain environmental standards, and I'm not going to do it. They will meet the standard like any other mill or be shut down" (personal interview).

In 1988, MB's plans to log the Carmanah Valley came under attack by environmental groups; MB responded by (unsuccessfully) seeking a court injunction to halt environmental groups from building trails in the area slated for logging. The BC government attempted to diffuse pressure over the Carmanah Valley by dividing it in two and permitting logging only in the upper half. The decision enraged environmental groups, and the

Nuu-Chah-Nulth Tribal Council threatened the BC government with court action. The Carmanah decision had also attracted the attention of a European environmental group that had previously organized a boycott of tropical hardwoods by European municipal governments. Much of the world watched as scrutiny focused on the region of Clayoquot Sound, a pristine area of old-growth forest on southwest Vancouver Island.

This time MB decided that it would need an environmental policy to help address all the environmentally related threats that it was facing, giving the environmental services department authority over the new policy. However, this department, primarily focused on technical issues concerning pollution control, was isolated from the actions of interest groups and had little understanding of issues associated with forest land use. It therefore had few capabilities to develop policies for addressing social stakeholder issues.[28] As the 1980s drew to a close, MB's dominant approach remained one of fending off or pacifying social pressures; the company believed that it could continue to rely on the provincial government to reduce these pressures. The 1989 annual report stated that "It is believed that the government recognizes the importance of a secure resource base to the company's ability to raise funds to invest in the very capital intensive pulp and paper sector of the province's forest industry. The Company believes that despite all the activity calculated to limit the industry in BC, the economic importance of forestry to the province will be recognized and loss of company cutting rights will be modest and largely compensable" (MacMillan Bloedel Ltd. 1990: 15-16).

The election of the NDP in 1991 marked the beginning of a new era of environmental regulatory change and a loosening of close government-forest industry ties.[29] Reflecting government policy, MB's timber supply was reduced to accommodate ecological concerns,[30] and an eighteen-month moratorium on several MB timber-harvesting areas was announced pending the outcome of new land-use initiatives that promised future withdrawals from the commercial land base. Now a multitude of MB forest management areas came under scrutiny, including the Walbran Valley (Wilson 1998).

On the pulp side, the spring and summer of 1991 saw a Greenpeace campaign in Europe aimed at convincing pulp buyers to use chlorine-free pulp. In January 1992, BC environment minister John Cashore announced that the province's environmental laws would be completely overhauled with extensive public consultation. New provincial pulp mill regulations to eliminate AOX in pulp mill effluent to undetectable amounts by 2002 were also announced. Federal regulations called for the reduction of dioxins and furans to nonmeasurable levels (actually a specific but very low level) by 1 January 1994 and extensive EEM monitoring on a three-year cycle. By 1992, MB acquiesced on pulp issues, investing capital in pollution control equipment to comply with regulations (though it was publicly critical of many of these new regulations).[31]

Protest in Clayoquot Sound

Amid all these changes, Clayoquot Sound became a lightning rod for international and domestic groups criticizing BC forest practices. After intensive cabinet debate, the provincial government announced in April 1993 that it would allow logging on two-thirds of Clayoquot Sound and remove one-third of it from commercial harvesting. MB supported the decision, though it noted the high price that it paid: the loss of some $40 million worth of finished products every year and 300 direct jobs. Environmental groups in general were outraged by the decision. The response by various environmental groups was to threaten blockades and civil disobedience. Some environmental leaders vowed to launch a major international campaign aimed at pressuring the provincial government to reconsider its decision. The media also speculated that the cabinet decision had revived the possibility of a European boycott of BC forest products. Environmental activists led by Greenpeace held around-the-world demonstrations to protest logging in Clayoquot Sound. The local group, Friends of Clayoquot Sound, launched a three-month campaign of civil disobedience that resulted in 1,000 protesters being arrested and 800 charged with criminal contempt.

MB's response to the blockades was initially to proceed as the company had in the past, adopting a public communications effort, attempting to consistently exercise its harvesting rights, taking legal action against individuals and groups,[32] and lending assistance to allies in the community.

However, new quasi-governmental institutions with no history of close industry-government collaboration made this traditional response more difficult. In particular, the Commission on Resources and the Environment (CORE), in its advisory capacity, analyzed the land-use decision and proposed the idea of a scientific panel to the government, which accepted this suggestion. In October 1993, the provincial government named its appointments to the Scientific Panel for Sustainable Forest Practices in Clayoquot Sound. Panel membership included biologists trained in ecosystem management, including Jerry Franklin, the architect of ecosystem management on US federal lands in the Pacific Northwest, whose management plans there had significantly reduced harvesting rates (Cashore 1999).

Meanwhile, a new Forest Practices Code was initiated as Greenpeace kicked off a worldwide campaign to convince MB customers in Europe and the United States to refuse to buy wood products originating from Clayoquot Sound. While the BC government used the code to say that things had changed (Bernstein and Cashore 2000; Cashore 1997), MB still took a fending-off approach. As one official stated, "Over four years we'll have spent in excess of $300 million. That's about 8 percent of our asset base, with no economic return to the company. We also have to pay the interest and added operating costs, so it's very expensive. The new BC regulations will increase that amount even further. Future approaches to environmental spending need to establish priorities intelligently"[33] (MacMillan Bloedel Ltd. 1991: 23).

Premier Mike Harcourt even criticized MB for not being proactive enough in defending itself against the boycotts, and MB quickly responded by sending a team of experts to Europe. Indeed, MB would spend about $1 million a year during this period on public relations. Yet these efforts did not derail the Greenpeace campaign, which met with considerable success. In December 1993, Greenpeace Germany convinced four of Germany's largest publishers to halt pulp and paper purchases from BC suppliers as soon as doing so was commercially feasible. In March 1994, the UK-based subsidiary of Scott Paper cancelled all contracts for pulp from forest companies operating in Clayoquot Sound after being pressured by Greenpeace.[34] In August 1994, the boycott campaign was extended to the United States when Greenpeace members sent letters to scores of US publishers and other buyers of pulp and paper produced in British Columbia, particularly by MB (Stanbury, Vertinsky, and Wilson 1995: 87).

MB appeared to have been caught off guard with respect to the Greenpeace campaign in Europe, and it did not have a proactive strategy in place. Economic stakeholders began to pressure MB. Dennis Fitzgerald of MB's public relations department described the response of MB customers who had been targeted by environmental groups: "Customers want to know, 'What the hell is going on? Why are people writing me these letters? Why are people picketing me outside my door, what is the whole list of charges Greenpeace is throwing at me, and most of all what are you guys doing to solve it? How are you going to make it go away?'" (personal interview). Fitzgerald observed that at the time there were no corporate answers. "We were outmatched, there's no doubt about it."

These events marked the first indication of a shift in MB's approach, moving slightly away from fending-off strategies and toward strategies of compromise. Two actions were key in this regard. First, MB took an active role in CSA forest certification development in an effort to gain recognition that the company was practising responsible forestry. Second, MB appointed a vice-president of environmental affairs almost as an ad hoc measure and began to publish an annual environmental report. The vice-president of environmental affairs at MB described her job as "unique":

I do environmental affairs, and it is a very unique position. You will not find another position like this in the forest industry here. I think that's probably a good thing for the other incumbents. Maybe my title isn't correct, I don't know. Mine is a policy position, so I am dealing with the company's broadly based policies on environmental performance and specific environmental issues ... My function has been very unique within it [MB]. I often describe my function as a corporate boutique function. It is not a line function to the extent that I don't have all these people working underneath me and reporting upwards or anything like that. It is a policy function. It is a little bit of a think-tank function. It is maybe an activist function. (personal interview)

Importantly, MB also began to find ways of compromising with environmental groups. MB and Greenpeace held secret meetings for the first time in October 1994. The dialogue was initiated by the Nuu-Chah-Nulth chiefs who had signed an interim measures agreement with the provincial government that gave them a voice in the use of resources at Clayoquot Sound. Apparently, MB was also under pressure from its customers to talk to Greenpeace. (The meetings broke off after Greenpeace blockaded GTE Directories, an MB customer in Los Angeles.) Although no agreement was reached, the meetings indicated that MB was considering a different approach.

In 1995, Greenpeace renewed its campaign against MB, taking its anti-clear-cutting message to a San Francisco convention of *Yellow Pages* publishers. Its strategy now was to target the entire BC coastal rainforest. It claimed that Clayoquot Sound was still being clear-cut even though regulations had suggested an end to clear-cutting. In May 1995, a coalition of environmental groups tabled a resolution at the annual general meeting of Pacific Telesis (the parent company of Pacific Bell) to end directory paper purchase from MB.[35]

The Greenpeace campaign was ongoing when the scientific panel reported to the provincial government, offering 127 recommendations based on implementing ecosystem management and extensive community involvement, including the use of historical Aboriginal harvesting techniques. The government accepted all of the panel's recommendations. MB had clearly lost the ability to rely on the government to act in its interests, as the recommendations would significantly reduce logging in Clayoquot Sound. When this became apparent, managers of the divisions in the sound were pulled out of MB's regular operating structure. An internal management board was created and chaired by the vice-president of environmental affairs to oversee operations there.[36] In January 1997, MB announced that it would downsize its Clayoquot Sound operations. The manager of environmental communications announced that MB had no plans to leave Clayoquot Sound, but there was no point in continuing operations on the same footing as in 1996, when MB lost $7 million on

Clayoquot Sound operations even though it was eligible for $3.2 million from Forest Renewal BC to help cover the cost of meeting the recommendations. "In 1996, we were lurching from little opening to little opening. We lost a bundle of money out there. It only harvested 52,000 cubic metres of a planned 100,000." However, the manager concluded that "The long-term picture may be much more optimistic than the short-term picture. Right now we've got a short-term position we have to deal with. We can't pretend we have enough work for those seventy-seven people" (personal interview).

The vice-president for environmental affairs announced that "MB does not believe its present operating structure in Clayoquot Sound works anymore" (personal interview). MB proposed shutting down all logging operations until sometime in 1998, stating that it wanted a clear end to uncertainty and a formal economic transition for local communities. Most environmental groups reacted favourably to the news, as MB was now on the defensive, looking for ways to find a compromise that would allow some degree of harvesting in the future.[37] Clayoquot Sound was a watershed for the company in the way that it dealt with environmental and Aboriginal groups, who used to be treated only as adversaries. Reflecting on these developments in 1997, the vice-president for environmental affairs said that the events at Clayoquot Sound were "a wake-up call ... It has had a seminal impact on the company's way of thinking. We've learned the hard way that the technical, scientific, factual, and economic answers don't represent the full equation anymore. There are social, political, and even philosophical and psychological dimensions to these issues ... The campaigns are mythic and emotional, and in a male-dominated, technical-based company people just didn't know how to deal with that" (personal interview).[38]

The new way of thinking paved the way for MB's announcement in April 1997 that the company would undertake a joint venture with Aboriginal groups in Clayoquot Sound.[39] A company official said that this venture was "an attempt to come to grips with some of these environmental and social issues in a way that also makes business sense" (personal interview). The agreement was seen as a first step by MB in converting all its Clayoquot Sound cutting rights to the joint venture. In May 1997, for the first time in years, MB's annual general meeting was remarkable for the absence of vocal protests from environmentalists.

Even more surprisingly, MB announced in April 1997 that it had entered into a joint venture with the Nuu-Chah-Nulth tribes to harvest 40,000 cubic metres a year of lumber from Clayoquot Sound starting in two years. The agreement had been the culmination of a year of negotiations. Environmental groups, though somewhat surprised, endorsed the venture. Greenpeace said that it had "trust in the First Nations vision," and the Western Canada Wilderness Committee (WCWC) said that it represented a "step toward sustainable management of the area." Both groups announced that they would not "twist the tail" of MB at its 1997 annual general meeting (personal interview).

These changes at MB can largely be attributed to its experiencing crises on several fronts. The actions of environmental groups precipitated a public relations crisis for the company. At the same time, MB experienced major problems with its institutional shareholders, who were critical of its financial performance.[40]

MB's approach toward environmental and Aboriginal groups had changed significantly from the early 1980s, as negotiations, discussions, and even joint ventures were deemed appropriate tools with which to address external pressures. Conceptions of sustainability were broadening, as the corporation recognized the need, at least in Clayoquot Sound, to accommodate a variety of interests and values in its forest operations. This approach paved the way for the next dramatic shift for MB.[41]

No Clear-Cutting in Old-Growth Forests
In June 1998, MB announced that it was developing a "forest project," which would include phasing out clear-cutting in old-growth forests, and initiating wide-ranging consultative decision-making processes.[42] Environmental pressures were no longer viewed as a constraint but an opportunity to be exploited on the market (MacMillan Bloedel Ltd. 1999). The company announced that it would begin to seek forest certification (green labelling) for its forests in order to gain social licence and to avoid the pressures from Europe that it had fought so persistently in the early to mid-1990s ("MacBlo" 1999).[43] The company was clear that, although it would first seek certification under the CSA forestry program, it would seek FSC certification once regional standards were agreed upon. MB also withdrew from the provincial industry's Forest Alliance, arguing that it had no interest in defending traditional harvesting methods (Hamilton 1998). Far from fighting environmental groups or even acquiescing to their demands, MB was now making a proactive response to environmental pressures. Now it was the BC government that urged caution, expressing concerns about timber supply and forest employment.

The turnaround from being one of the most intransigent companies to being one of the most proactive can be traced to the new approach that MB began to take in Clayoquot Sound and to the hiring of a new CEO, Tom Stephens. When he first arrived, it was unclear whether the position of vice-president for environmental affairs would be retained. However, after a comprehensive review of company operations, Stephens became convinced that increasing the company's social licence could also help its bottom line.[44] One means to improving social licence was an agreement with Greenpeace that it would stop its boycott campaign, and in return MB would promise to phase out clear-cutting in its old-growth forests and to introduce no-harvesting and low-harvesting impact zones (MacMillan Bloedel Ltd. 1998a).[45] An implementation advisory committee would be constituted and include experts nominated by environmental and Aboriginal groups.[46]

When the announcement was made, the unthinkable happened as Greenpeace officials gave Stephens a bottle of champagne and congratulated him on the reversal of policy (Hogben and Canadian Press 1998). MB was still acting as a profit-maximizing firm, but now the ecological issues were seen as an opportunity rather than a constraint. As Stephens explained to shareholders at MB's 1999 annual general meeting, "we want to move from reactive to proactive and be the creator and driver of trends so that MB is the winner. This calls for leadership and being a *trend setter* not just a *trend follower*" (Stephens 1999).

The incentive to change came from a combination of boycott pressures and the growth of forest certification schemes internationally and domestically (Bernstein and Cashore forthcoming). MB's conception of sustainability had moved closer toward the more holistic orientation.

This did not mean that MB would end all hostilities with environmental groups. The company also came under fire from BC environmental groups in April 1999 for a deal with the provincial government to transfer ownership of some of its areas under licence to the company in exchange for dropping its litigation demands that it be compensated for the government's forest preservation initiatives. Environmental groups were particularly concerned because transferring land to MB would remove it from the scrutiny of the environmental provisions of the new Forest Practices Code (Armstrong 1999).

At the same time, MB has been diversifying beyond British Columbia and selling many of its assets in the province (Gibbon 1999). Indeed, MB returned to the black in the first quarter of 1999, largely as a result of profits from its private forests, which were not subject to its new clear-cutting rules (Hunter 1999). In addition, MB continues to be critical of the government's environmental forest policies. CEO Stephens (1999) stated that the company will not invest more in British Columbia until the province gives in more on regulations, arguing that "logging on Crown land under today's rules is just not something we are willing to invest money in because the risk and the reward are out of balance."

Lessons from the Case Studies
The three cases outlined above reveal a complex story in which firms respond to environmental groups, economic stakeholders/pressures, and governmental agencies/rules in different ways at different times (Table 4.2). While all three companies dismissed or pacified external pressures

from the mid-1970s to the mid-1980s, they followed different paths after this period. They reveal that firms are susceptible to changes in how they address external pressures, rendering long-lasting characterizations difficult. Indeed, MB was the most reluctant to address societal criticisms and environmental pressures, but when it decided to change course it went beyond anything that Canfor or Weyerhaeuser had proposed. Moreover, as these companies sought accommodating and proactive solutions in the late 1990s to environmental groups and societal pressures, their responses were decidedly different from those to governmental pressures. Notably, Canfor and MB became increasingly hostile toward provincial environmental policies, arguing that an array of policies, including the Forest Practices Code, stifled a competitive climate. This reaction occurred when they were proactively embracing certification programs designed to address forest practices in the private sphere. The positive social licence that came with supporting forest certification and the negative boycott campaigns that would have materialized had MB and Canfor not supported certification appear to have been key reasons for their support of private sector forest practices rules. However, such incentives did not apply to governmental policies, and the firms thus took a different approach in this arena.

While important generalizations can be made, the cases also reveal the difficulty in categorizing every firm's actions within our classification scheme. The cases highlight the constant need to understand the contexts within which choices are made. For example, Weyerhaeuser's efforts to avoid endangered species rules arising out of the Endangered Species Act through the development of alternative habitat conservation plans can be seen as both *avoiding* and *proactive/innovative*. The firm was clearly motivated to avoid what would have been economically more difficult rules by creating an innovative conservation plan that it argued would allow more harvesting without compromising species recovery efforts.

Likewise, a distinction must be made between responses to pulp mill pollution issues and forest practices matters. In the cases above, these large companies had to respond to pressures in each area, but notably a greater degree of Canfor's pressure came on the pollution front, while criticisms of MB were largest for forest practices/old-growth issues. Indeed, the cases reveal a shift in environmental group and societal concerns from a focus on water and air pollution in the late 1980s to forest practices in the 1990s. Until the 1990s, in fact, Greenpeace had never been involved in BC forest practices matters (see Cashore 1997).

Given this complexity, which conclusions can we draw about the nature of a firm's responses to external pressures? Can governments and civil society rely on corporations to implement sustainability initiatives? What does the review tell us about the roles of environmental groups, governmental

Table 4.2

Company responses to external pressures

	Mid-70s to mid-80s	1987	Late 1989	1993	1998
Weyerhaeuser	Dismiss, pacify	Dismiss, pacify	Accommodate Proactive	Accommodate Proactive	Accommodate Proactive (citizen and environmental groups) Compromise/fend off (governmental agencies)
MacMillan Bloedel	Dismiss	Dismiss	Dismiss, defy, fend off	Acquiesce, compromise	Accommodate Proactive/innovative (to environmental groups) Fend off (government)
Canfor	Dismiss, pacify	Manipulate, defy	Accommodate Proactive/ innovative	Accommodate Proactive (to environmental groups) Compromise (government)	Accommodate Proactive (to environmental groups) Fend off (government)

policy, and a company's place in the global economy in influencing firm-level responses to pressures for sustainable forest management? The following propositions summarize the main conclusions drawn from the case studies.

(1) *When an issue becomes highly publicized, companies will act for fear of losing "social licence."* In each of the cases above, forest companies eventually responded to pressures when they were losing favour with the public and when the lack of corporate response had led the issue to enter what Cobb and Elder (1972) label the "systemic" (societal) policy agenda. While social licence is hard to measure, the uncertainty associated with its loss appears to be something that industry officials wish to avoid (Reed 1999). At the same time, the desire for social licence is not, by itself, sufficient to trigger changes in corporate responses to external pressures.

(2) *Environmental group pressures are most effective when accompanied by economic pressures for change.* In the three cases above, company officials decided either to acquiesce or to take proactive measures when they believed that not doing so would hurt corporate bottom lines. In none of the cases did a firm act simply because of normative value changes. Do firms that first act with proactive and innovative responses eventually undergo normative value changes? The hiring of new officials from different backgrounds suggests that new values will enter the firm, a process that one official referred to as being "infected." The issue is whether this infection lasts, even when economic pressure no longer exists.

(3) *Firm-level changes are limited to "free-market environmentalism."* From the three cases above, it appears that profit-maximizing firms will only find solutions to environmental pressures that also allow them to develop (at least potentially) new markets or ultimately improve their bottom lines. This statement may be rather obvious given that firms operate within the constraints and incentives of a capitalist economy. But acknowledging it leads to the recognition that environmental choices are limited to the logic of capitalism and economic growth. If both capitalism and economic growth are deemed to be unsustainable paradigms, then it is highly unlikely that a solution can be found at the level of the firm, because it is in the interest of a firm to maintain the capitalist environment in which it operates.[47]

Arguably, the "crown jewel" for environmental groups in British Columbia has been Clayoquot Sound, in which logging has been reduced to a fraction of what it once was and an inclusive ecosystem management approach has been adopted. But it is ironic that MB has virtually written off Clayoquot Sound as a profitable enterprise, treating it instead as an innovative experiment in an alternative form of harvesting. Many analysts argue that it is doubtful that MB or any large integrated forest company could survive profitably if Clayoquot rules were applied across the province. Indeed, when MB returned to the black in 1998, analysts revealed that this profit was entirely due to MB's private land operations, which have been relatively shielded from public scrutiny (Hunter 1999).

(4) *Limited and discretionary regulations allow for innovation but also encourage noncompliance.* All three firms undertook proactive or acquiescent responses to environmental pressures while promoting limited regulations. Indeed, many of Weyerhaeuser's responses were facilitated by a regulatory regime that allowed the company to escape requirements of the Endangered Species Act if it could come up with an equivalent but less burdensome alternative. Canfor and MB both supported the idea of a Forest Practices Code but were critical of its complicated and onerous regulatory requirements (though many of the rules were discretionary). The evidence here supports Vertinsky and Zietsma's (1998) claim that "greening" and "innovation" tend to occur when firms are not burdened with difficult and bureaucratic regulatory requirements.[48]

Caution is warranted, however. A limited regulatory regime has also been shown to encourage noncompliance. There is evidence that a nonlegal/discretionary regulatory style alongside a clientele/pluralist system of network governance results in poor compliance with existing state regulations (Environment Canada 1998; Tripp, Nixon, and Dunlop 1992). At the same time, other studies have found that the highest form of corporate compliance is achieved with a strict nondiscretionary legal regime. Jennings and Zandbergen's literature review (1995: 1023-4) found broad support for the notion that firms will acquiesce when institutional pressures are in the form of legal coercion, but at the same time firms are unlikely to undertake normative or mimetic responses under such conditions. The lessons for regulators appear to be that a limited regulatory regime facilitates both proaction and noncompliance, while a stronger nondiscretionary regulatory regime promotes acquiescence.[49]

(5) *A firm's conception of sustainability affects its choices in responding to external pressures.* All three firms responded to environmental pressures by acquiescing or taking proactive measures once their definitions of sustainability had moved from a focus on timber toward a more holistic conception. Indeed, MB's decision to react negatively to pressures for such a sustained period appears to have been related to its underlying understanding of forest sustainability and its steadfast belief that the company was practising scientifically sound sustainable

forestry. Weyerhaeuser and Canfor also had similar initial responses, though their conceptions of the term changed far more quickly than MB's did. Just how deep these conceptions are remains a question for future investigation. If a firm can change quickly to one conception, then can it also move quickly away from such a conception if external scrutiny subsides?

(6) *A firm's organizational structure matters.* In our three cases, Weyerhaeuser and Canfor, the firms with team-oriented, less hierarchical organizational structures, were the quickest to adapt to new pressures, while MB's hierarchical, competitive atmosphere revealed a slow capacity to adapt. Indeed, it was not until MB changed its organizational structure (by adding the position of vice-president for environmental affairs outside the hierarchical structure) that it adapted more easily and considered making innovative responses.

(7) *A new CEO makes a difference.* New CEOs can be the sources of new ideas and change. Weyerhaeuser changed its response quickly after it hired a new CEO in the late 1980s, and MB's hiring of a new CEO preceded its significant turnaround. At the same time, CEOs are constrained by past activities and the institutional environments in which they operate.

(8) *Land ownership is important.* Weyerhaeuser's choice to keep its forest land in the early 1900s facilitated its stewardship approach, first in adopting timber sustainability and later in expanding this definition toward the more holistic conception. Private forest land ownership may encourage the first view of sustainable forest management, while public forest land ownership may encourage the second view.[50]

Conclusion

Weyerhaeuser's Purchase of MacMillan Bloedel

This chapter began by arguing that the dual forces of economic globalization and internationalization (i.e., external forces that influence domestic policies) are increasingly important in understanding firm-level policy choices. Perhaps the best illustration of these forces is that the three firm-level cases chosen for this chapter have been reduced to two as Weyerhaeuser's efforts to expand and diversify resulted in its purchase of MB. This acquisition raises issues and questions. It reveals the difficulty in attempting to categorize a single firm, which may have different cultures and approaches within the operation. Will the old MB operations be different from Weyerhaeuser's operations in the US Pacific Northwest? Or will a single identifiable company emerge?

Somewhat ironically, MB's 1998 forest project meant that the company had become a corporate favourite of environmental groups, and now there was concern that Weyerhaeuser might backtrack on MB's certification and old-growth commitments. These fears were heightened when, on the day of the acquisition announcement, Weyerhaeuser officials refused to commit to MB's forest project, explaining that they would first need to study the situation (Barrett and Hunter 1999). Environmental groups were quick to fight the proposed purchase, ironically arguing that MB was now further advanced than Weyerhaeuser on environmental issues (David Suzuki Foundation 1999). However, it seems unlikely that Weyerhaeuser will backtrack, because such a move would mark a change from the approach that it has taken in the US Pacific Northwest. Indeed, the official report created to advise the BC government on allowing the purchase or not noted that, whether the company is owned by MB or Weyerhaeuser, the international market pressures that spurred MB's forest project are still there, and the consequences of changing that project could hurt the corporate image, reduce social licence, and hinder European markets (Perry 1999: 48).

Concepts of Sustainability

This chapter started by identifying two concepts of sustainable forest management. While none of the companies discussed here rejected a timber sustainability approach or embraced completely a more holistic concept, a decision to undertake innovative or proactive responses to external pressures was preceded by a move away from the predominance of a timber sustainability concept. The case studies reveal that a host of new pressures has entered forestry decision making, including pressures from environmental groups, the effects of economic globalization, boycott threats by customers, and international scrutiny. These pressures affect firms in different ways and mould their relationships and responses to government regulations. Understanding these relationships is key to the design of effective regulatory strategies. Furthermore, as governments are increasingly under pressure to reduce their scope yet increase environmental protection, it is important to examine the ways in which corporate greening might develop without direct government interaction.

At the same time, more research needs to be done to examine the relationship between changes in corporate responses and on-the-ground environmental performance. Do companies that have undertaken proactive responses reduce the impacts of their operations more than companies that take a defiant approach to environmental issues? Is there an overall cost or benefit to taking proactive responses? We noted above that in key areas such as Clayoquot Sound profitable forest operations have been, at least in the short term, written off. In these cases, proactive responses meant curtailing economic activity. The benefit of social licence outweighed the loss in profits, largely because Clayoquot Sound was a small fraction of MB's operations. Such an approach company-wide would result in a significant reduction in company operations or simply going out of business, a result that would undermine the contention that corporate greening is good for both the environment and the bottom line (Elkington 1998).

Clearly, more research has to be undertaken into the complex relationship between social and economic systems and their effects on the natural environment. This chapter is a small but important component within this larger project. While voluntary firm-level sustainable initiatives are clearly not a panacea for environmental protection,[51] they are an important piece of the puzzle for those who wish to understand how a sustainable economy might be achieved and the opportunities and obstacles that will be encountered.

Notes

1 The sections on globalization and internationalization draw heavily on Benjamin Cashore's collaboration with Steven Bernstein (see citations below). The authors are grateful to George Weyerhaeuser Jr., Bob Prolman, and an anonymous reviewer for detailed and thoughtful comments on an earlier version of this chapter. The authors also thank Michelle Vasser for helpful research assistance. Financial support for this research was provided by the National Centres of Excellence, Sustainable Forestry Program; Industry, Trade and Economics, Canadian Forest Service – Pacific Forestry Centre; Natural Resources Canada; the University of British Columbia's Hampton Fund; and the Alabama Agricultural Experiment Station.

2 Weyerhaeuser has significant holdings in Canada, but this analysis focuses primarily on its US operations.

3 The concept of "social licence" or "social licence to operate" (SLO) is now gaining attention within the forest policy community and is being used by officials at MacMillan Bloedel in British Columbia to explain their forest project. The implications and dynamics of SLO are undergoing research at the World Resources Institute (WRI) (see Reed 1999). Aside from the WRI project, a literature search revealed no published scholarly articles using this term.

4 See Bernstein and Cashore (1999); Cashore (1995a, 1995b, 1995c, 1997, 1999); Hoberg (1993, 1996a, 1996b); Hoberg and Morawski (1997); Lertzman, Wilson, and Raynor (1996a, 1996b); Wilson (1998); and Yaffee (1994).

5 Notable exceptions include Cashore and Vertinsky (2000) and Day, Hart, and Milstein (1998).

6 Oliver also offers a range of tactics that a company might choose under each strategy that it adopts. Manipulation involves tactics to co-opt, influence, or control external pressures; defiance involves dismissing, challenging, or attacking; avoidance results in concealing, buffering, or escaping; compromise involves balancing, pacifying, or bargaining; and acquiescence entails imitating or complying.

7 Bernstein (forthcoming a and b).

8 The commission defined sustainable development as "development that meets the needs of the present without compromising the ability of future generations to meet their own needs (World Commission on Environment and Development 1987: 41).

9 Lesser known programs are also emerging as possible certification alternatives in the United States, including the American Tree Farm System established by the American Forest Foundation and the Green Tag program established by the National Forestry Association.

10 See Forest Stewardship Council (1999b) and Moffat (1998: 44).

11 Final approval from Oaxaca for US south FSC standards is pending (Forest Management Trust 1999).

12 Environmental groups include Natural Resources Defense Council, World Wildlife Fund, Greenpeace, Sierra Club, Wilderness Society, National Wildlife Federation, Friends of the Earth, Environmental Defense Fund, Rainforest Action Network, Rainforest Alliance, World Resources Institute, American Lands Alliance, Ecotrust, Institute for Sustainable Forestry, and others.

13 Cited in Hansen and Juslin (1999).

14 Regeneration is usually undertaken not for environmental concerns but because industry has an economic self-interest in creating fast-growing fibre. As the AF&PA notes, "The US forest and paper industry understands this concern. Companies that rely on healthy and productive forestland for their livelihood have a keen self-interest in making certain that US forests remain healthy and productive" (1995: 4).

15 To date, Champion, Meade, International Paper, and Plum Creek have opted for third-party verification.

16 Semi-structured personal interviews were conducted with key informants between October 1996 and March 1997. The key informants were selected from among upper and middle management with the objective of obtaining representation from all the corporate departments and divisions with responsibility for the development of environmental policy and related initiatives. An effort was made to interview comparable informants (with respect to position within the company) at the participating companies. The interviews, which lasted between 1-2 hours, were recorded and transcribed. Informants were then sent a copy of the transcript with the opportunity to review and corroborate their responses and provide any additional comments.

17 The ESA allows private landowners to develop a "habitat conservation plan" (HCP) that the secretary of interior considers for approval. The HCP must include a list of possible impacts of an action (e.g., logging), steps to be taken to limit detrimental effects, and a justification of the plan over other options (Smith, Moote, and Schwalbe 1993: 1039). If an HCP is approved, then an "incidental take permit" is issued for a project. See Chapter 6 in this volume.

18 See Durbin and Koberstein (1990) and Salazar (1988).

19 Also see Hoberg in Chapter 2 of this volume.

20 Raizada (1998: 163) notes that Canfor argued that it had a long-established commitment to reforestation and wildlife conservation.

21 The *Vancouver Sun* of 14 December 1987 reported that the company had never been charged with its failure to comply with the original 1978 BC waste management permit for the mill despite consistently exceeding agreed-upon levels.

22 Both Greenpeace and the US Environmental Protection Agency released data in 1987 showing that pulp mill effluent contained traces of dioxin, a toxic organochlorine (Raizada 1998). Canfor's pulp mill operations were now part of a general provincewide antieffluent campaign that included Greenpeace Canada, the BC-based Sierra Club, the West Coast Environmental Law Association, and the Western Canada Wilderness Committee, which were part of a larger international effort by Greenpeace and the World Wildlife Fund in Europe (Raizada 1998: 167). The campaign was boosted by a 1988 study showing that most pulp and paper mills in British Columbia were not in compliance with 1971 federal pulp and paper effluent regulations.

23 The regulations would focus on organochlorines in pulp mill effluent (measured through AOX).

24 International attention culminated at the 1992 Earth Summit in Rio de Janeiro, which also helped to pave the way for the creation of forestry certification schemes such as that of the Forest Stewardship Council (Bernstein and Cashore forthcoming).

25 Quoted in Raizada (1998: 188). The Howe Sound mill became the first kraft mill in North America to complete successfully a full-scale trial of chlorine-free bleached softwood kraft market pulp.

26 MB's annual report for 1985 included a section on reforestation entitled "The British Columbia Forest: A Renewable Resource." Approximately 58 percent of MB's forest land was classified as old-growth mature timber, 39 percent was young forest, and 3 percent was not satisfactorily restocked (NSR) land.

27   Also criticized was MB's clear-cutting of these forests, though preservation rather than sustainable forestry was the clear focus of these campaigns. Somewhat ironic was a secondary complaint about the practice of "high-grading," in which MB would only log trees that provided the most value and leave behind timber of lesser value. As we will show later, MB would successfully alter the debate from old-growth preservation to the elimination of clear-cutting in old-growth forests – moving toward another form of selective logging.

28   The environmental policy focused on four areas: complying with laws and regulations, minimizing the risk of hazardous events, undertaking regular auditing, and ensuring that managers and supervisors complied with the policy. As early as 1989, MB's own Public Affairs Group had developed an action plan that called for sweeping changes in MB's forest practices at an annual cost of $50 million. The plan was rejected by MB's senior management on the grounds that it did not make a strong enough case that the changes would benefit the environment or give MB any competitive advantage.

29   Almost half of the promises in the party's fifty-four-point platform dealt with natural resources and/or environmental issues. A key commitment was to double the area devoted to parks or ecological reserves to 12 percent of the province's land base by the year 2000. This promise followed the Brundtland report's recommendation that 12 percent of the world's land be protected from development.

30   In January 1992, the chief forester scaled back cutting rights on southern Vancouver Island in three tree farm licences, including MB's, as part of the ongoing provincial timber supply review (a process to reevaluate the sustainability of allowable annual cuts).

31   However, pulp issues again became a concern after a chlorine dioxide spill at its Powell River mill in 1994. MB potentially faced fines of more than $1 million resulting from charges laid in connection with the hazardous spill. The board, largely stimulated by its increased personal liability for environmental disasters, responded by directing the environmental services department to develop a corporate chemical process safety system management standard (completed by 1996).

32   MB announced during the blockades that it would sue Greenpeace for damages.

33   Most of the money spent by MB went into pollution control for mills and not to forestry operations.

34   MB sold about 3 percent of its market pulp to Scott Limited.

35   This was the climax of an escalating ecocampaign in California against Pacific Bell, which buys up to 35 percent of its directory paper from MB. Greenpeace said that MB had been singled out because it continued to clear-cut in Clayoquot Sound. Representatives of the federal Canadian and provincial BC governments, MB, and the Forest Alliance went to San Francisco to deal with customers' concerns.

36   The management structure that already existed there was left in place, and it still reported up through the normal woodlands structure. A dual line of authority was created, the second one going up to the environmental affairs department.

37   Greenpeace, Rainforest Action Network, and Friends of Clayoquot Sound all adopted a wait-and-see stand and the temporary cessation of actions against the company. Greenpeace said that it was counting on Clayoquot Sound being declared a UN biosphere reserve, which would free up federal dollars to assist loggers to make the transition to other work.

38   An MB manager commented how some initiatives, such as the publication of an annual environmental report, had been instrumental in bringing people together from different parts of the organization to discuss environmental issues and public perceptions of MB's responsiveness to these issues.

39   The Aboriginal groups would own 51 percent of the venture under the auspices of the Ma'Mook Development Corporation. MB would turn control of the northern half of the forests of Clayoquot Sound over to the venture. MB said that it was a way to heal past conflicts in the region.

40   This criticism provoked a major restructuring. The expensive corporate office was downsized, and separate business units were set up to run its paper operations and its building materials divisions from its research facility in Burnaby. MB sold its shipping division and announced that it would trim the ranks of its vice-presidents.

41 This shift, while important, should not be overstated. As a profit-maximizing corpora-
tion, MB had serious concerns about its ability to continue to operate in the province.
During this time, CEO Robert Findlay argued that "MB's days of growth in BC are over ...
There are no longer opportunities for MB to grow in BC. Our AAC is receding, so if we
are going to grow as a company, we are going to grow somewhere else" (*Vancouver Sun,* 1
May 1997: D6).

42 See Hamilton (1998); Hunter (1998); and MacMillan Bloedel Ltd. (1998a, 1998b).

43 On 14 April 1999, the company announced that its North Island Woodlands Division on
Vancouver Island "had passed an independent audit on the way to becoming the first
operation in Canada certified to the Canadian Standards Association's (CSA) Sustainable
Forest Management Standard" (MacMillan Bloedel Ltd. 1999) and one of the first in
North America to be certified under the International Organization for Standardization
(ISO) 14001 environmental management systems program. That MB started first with
CSA and ISO did not mean that it was ignoring the Forest Stewardship Council (FSC) program, which has much greater support from the domestic and international environmental communities (Bernstein and Cashore 1999). MB noted in the same press release
that "The Forest Stewardship Council system is definitely of interest to some of our customers, but we've deferred pursuing that to allow time for the development of FSC
regional standards in BC. Our practical experience in implementing the ISO and CSA
standards will allow us to contribute productively to the FSC work in this province"
(MacMillan Bloedel Ltd. 1999). See also Matas and Lush (1998). Indeed, the company
asked environmental groups to give MB their views on the FSC certification standards for
British Columbia (e-mail from MB posted on commforest@onenw.org, 22 April 1999).

44 Many have noted the unique ability of Vice-President Linda Coady to interact with representatives of environmental groups. Most of the Greenpeace officials were women and
came from social science backgrounds. Indeed, Coady and Greenpeace officials lived in
the same neighbourhood and often discussed issues while walking their babies, a phe-
nomenon labelled "baby buggy diplomacy" ("Greening" 1998).

45 By negotiating with Greenpeace, MB could refocus the debate away from preservation of
old-growth forests and toward simply not clear-cutting them. This shift is important,
because groups focused on clear-cutting rather than preservation during the Clayoquot
Sound crisis in order to appeal to Aboriginal interests. Indeed, Greenpeace's Vancouver
officer ran a newspaper ad in the *Globe and Mail* in 1994 that focused on clear-cutting
rather than preservation (Greenpeace Vancouver 1994).

46 The myriad of sustainable forestry initiatives was undertaken under the auspices of the
MB forest project. The title of vice-president for environmental issues was changed to
"environmental enterprises." As the holder of this position, Linda Coady oversaw the forest project, particularly with respect to MB's operations in coastal forests (see MacMillan
Bloedel Ltd. and Dovetail Consulting Inc. 1999).

47 Bernstein (forthcoming) argues that a pro-market/growth "liberal environmentalism"
approach has dominated international environmental politics.

48 The issue of whether a proactive firm can be categorized as "green" is controversial
because a proactive firm in a limited regulatory climate might actually do less for the
environment than an "acquiescent" firm in a highly regulated climate.

49 For a detailed discussion, see Cashore and Vertinsky (2000).

50 See Burda, Gale, and M'Gonigle (1998).

51 For a discussion of the problems with a firm's adopting market-oriented certification
schemes, see Gale and Burda (1997).

References

American Forest and Paper Association. 1995. *Sustainable Forestry Initiative.* Washington,
DC: American Forest and Paper Association.

Armstrong, Jane. 1999. "BC Deal with Loggers Discards Environmental Rules: Province
Hands Over Land in 'Horrible Precedent.'" *Globe and Mail* 18 March, Internet posting,
www.globeandmail.ca.

Barrett, Tom, and Justine Hunter. 1999. "US Giant Urged to Honour Environmental Deals." *Vancouver Sun* 24 June: A3.

Batsell, Jake. 1998. "Weyerhaeuser Closing Oregon Sawmill." *Seattle Times* 13 January: F11.

Bernstein, Steven. 1998. "Logjam in Global Forest Protection: Contesting Liberal Environmentalism." Paper presented to the Thirty-Ninth Annual Convention of the International Studies Association, Minneapolis.

—. Forthcoming a. "Ideas, Social Structure, and the Compromise of Liberal Environmentalism." *European Journal of International Relations* 6 (4).

—. Forthcoming b. *The Compromise of Liberal Environmentalism*. New York: Columbia University Press.

Bernstein, Steven, and Benjamin Cashore. 1999. "World Trends and Canadian Forest Policy: Trade, International Institutions, Consumers, and Transnational Environmentalism." *Forestry Chronicle* 75 (1): 34-8.

—. 2000. "Globalization, Fourth Paths of Internationalization, and Domestic Policy Change: The Case of Eco-Forestry Policy Change in British Columbia, Canada." *Canadian Journal of Political Science* xxiii (1) (March): 67-99.

—. Forthcoming. "The International-Domestic Nexus: The Effects of International Trade and Environmental Politics on the Canadian Forest Sector." In Michael Howlett (ed.), *Canadian Forest Policy: Regimes, Policy Dynamics, and Institutional Adaptations*. Toronto: University of Toronto Press.

Brown, Leslie. 1997. "Weyerhaeuser Plans Overseas Timberland Buy." *Journal of Commerce* 17 January: 11B.

Bruce, Robert A. 1998. "The Comparison of the FSC Forest Certification and ISO Environmental Management Schemes and Their Impact on a Small Retail Business." MBA thesis, University of Edinburgh.

Burda, Cheri, Fred Gale, and Michael M'Gonigle. 1998. "Eco-Forestry versus the State(us) Quo: Or Why Innovative Forestry Is Neither Contemplated Nor Permitted within the State Structure of British Columbia." *BC Studies* 119: 45-72.

Canfor Corporation. 1988. *Canfor Overview*. April. Vancouver: Canfor Corporation.

—. 1992. *Annual Report: Canfor Corporation*. Vancouver: Canfor Corporation.

—. 1998. *Annual Information Form*. Vancouver: Canfor Corporation.

Cashore, Benjamin. 1995a. "Comparing the Eco-Forest Policy Communities of British Columbia and the US Pacific Northwest." Paper presented at the Biennial Meeting of the Association for Canadian Studies in the United States, Seattle, 15-19 November.

—. 1995b. "Comparing the Eco-Forest Policy Regimes of British Columbia and the US Pacific Northwest." Paper presented at the Canadian Political Science Association Annual Meeting, Montreal, 4-6 June.

—. 1995c. "Explaining Forest Practice and Land Use Policy Network Divergence in British Columbia and the US Pacific Northwest." Paper presented at the Pacific Northwest Political Science Association Annual Meeting, 19-21 October, Bellingham, WA.

—. 1997. "Governing Forestry: Environmental Group Influence in British Columbia and the US Pacific Northwest." PhD diss., University of Toronto.

—. 1999. "US Pacific Northwest." In Bill Wilson et al. (eds.), *Forest Policy: International Case Studies*. 47-80. Oxon, UK: CABI Publications.

Cashore, Benjamin, and Ilan Vertinsky. 2000. "Policy Networks and Firm Behaviours: Governance Systems and Firm Responses to External Demands for Sustainable Forest Management." *Policy Sciences* 33: 1-30.

Cobb, Roger W., and Charles D. Elder. 1972. *Participation in American Politics: The Dynamics of Agenda-Building*. Boston: Allyn and Bacon.

"Concern Seeks to Build Fund to Buy Foreign Timberlands." 1995. *Wall Street Journal* 14 April: C11.

David Suzuki Foundation. 1999. *MacBlo-Weyerhaeuser Deal Bad for BC's Forests*. News release. 22 June 1999.

Day, Robert, Stuart Hart, and Mark Milstein. 1998. *Weyerhaeuser Forestry: The Wall of Wood. Case Study*. Covelo, CA: Island Press.

DiMaggio, Paul J., and Walter W. Powell. 1991. "The Iron Cage Revisited: Institutional

Isomorphism and Collective Rationality." In Walter W. Powell and Paul J. DiMaggio (eds.), *The New Institutionalism in Organizational Analysis*. 63-82. Chicago: University of Chicago Press.

Drushka, Ken. 1995. *HR: A Biography of H.R. MacMillan*. Madeira Park, BC: Harbour Publishing.

Durbin, Kathie, and Paul Koberstein. 1990. "Forests in Distress: Special Report." *Oregonian* 15 October: 1-28.

Elkington, John. 1998. *Cannibals with Forks: The Triple Bottom Line of the 21st Century Business*. Gabriola Island, BC: New Society Publishers.

Environment Canada. 1998. *Enforcement versus Voluntary Compliance: An Examination of the Strategic Enforcement Initiatives Implemented by the Pacific Yukon Regional Office of Environment Canada*. Ottawa: Environment Canada.

The Forest Management Trust. 1999. *Forest Certification Handbook: For the Southeastern United States*. Gainesville, FL: Forest Management Trust.

Forest Stewardship Council. 1999a. *Forests Certified by FSC-Accredited Certification Bodies*. Oaxaca, Mexico: Forest Stewardship Council.

—. 1999b. *FSC Principles and Criteria*. Oaxaca, Mexico: Forest Stewardship Council.

Gale, Fred, and Cheri Burda. 1997. "The Pitfalls and Potential of Eco-Certification as a Market Incentive for Sustainable Forest Management." In Chris Tollefson (ed.), *The Wealth of Forests: Markets, Regulation, and Sustainable Forestry*. 414-41. Vancouver: UBC Press.

Gibbon, Ann. 1999. "Alliance Forest Mulls Takeover Bid for Pacifica: Weighs $400-Million Offer for MacBlo Chapter Spinoff." *Globe and Mail* 19 January, Internet posting, www.globeandmail.ca.

"The Greening of MacMillan Bloedel." 1998. Advertisement. *Adbusters* June: 54-5.

"Greenpeace turns up pressure on Germany." 1993. *Globe and Mail* 9 July: B3.

Greenpeace Vancouver. 1994. "Wanted: Clear-cut-Free Forest Products." Advertisement. *Globe and Mail*: B4.

Hall, Terry. 1997. "Weyerhaeuser Buys New Zealand Forest." *Financial Times*, London Edition, 12 April.

Hamilton, Gordon. 1998. "MacBlo Decides to Abandon Pro-Logging Forest Alliance." *Vancouver Sun* 18 August, Internet posting, www.vancouversun.com.

Hansen, Eric, and Heikki Juslin. 1999. *The Status of Forest Certification in the ECE Region*: 42. New York and Geneva: United Nations, Timber Section, Trade Division, UN-Economic Commission for Europe.

Hoberg, George. 1993. *Regulating Forestry: A Comparison of Institutions and Policies in British Columbia and the US Pacific Northwest*. Vancouver: Forest Economics and Policy Analysis Research Unit, University of British Columbia.

—. 1996a. "The Politics of Sustainability: Forest Policy in British Columbia." In Ken Carty (ed.), *Politics, Policy, and Government in British Columbia*. 272-89. Vancouver: UBC Press.

—. 1996b. "Putting Ideas in Their Place: A Response to 'Learning and Change in the British Columbia Forest Policy Sector.'" *Canadian Journal of Political Science* 29(1): 135-44.

Hoberg, George, and Edward Morawski. 1997. "Policy Change through Sector Intersection: Forest and Aboriginal Policy in Clayoquot Sound," *Canadian Public Administration* 40(3): 387-414.

Hogben, David, and Canadian Press. 1998. "Environmentalists Toast MB's Clear-Cut Decision." *Vancouver Sun* 11 June: D1, D8.

Hunter, Justine. 1998. "MacBlo Decides to End Clear-cutting in Old-Growth Coast Forests: The Company Is Trying to Counter Damage to Its Sales Because of Its Environmental Critics." *Vancouver Sun* 10 June: A1, A2.

—. 1999. "MacBlo Leads the Parade Back into the Black." *Vancouver Sun* 22 April, Internet posting: www.vancouversun.com.

Hunter, Justine, and David Hogben. 1998. "Stumpage Cuts Trigger Rehiring in BC Forests: The Premier's Initiative Brings Negative Reaction from US Lumber Industry." *Vancouver Sun* 29 May: A1-2.

International Development Research Centre. 1993. *The Earth Summit*. CD-ROM. Ottawa: International Development Research Centre for the United Nations.

Jennings, P. Devereaux, and Paul A. Zandbergen. 1995. "Ecologically Sustainable Organi-
zations: An Institutional Approach." *Academy of Management Review* 20,4: 1015-52.
Johnson, Nels. 1993. "Introduction to Part I, Sustain What? Exploring the Objectives of
Sustainable Forestry." In Greg Aplet et al. (eds.), *Defining Sustainable Forestry*. 11-15.
Washington, DC: Island Press.
Lambert, Cheryl Ann. 1996. "Weyerhaeuser Pays for Water Violations." *Home Improve-
ment Market* September.
Lertzman, Ken, Jeremy Wilson, and Jeremy Raynor. 1996a. "Learning and Change in the
B.C. Forest Policy Sector: A Consideration of Sabatier's Advocacy Coalition Framework."
Canadian Journal of Political Science 29(1): 112-33.
—. 1996b. "On the Place of Ideas: A Reply to George Hoberg." *Canadian Journal of Political
Science* 29(1): 145-48.
"MacBlo Courting Environmental Approval." 1999. *Canadian Press* 24 February, Internet
posting, www.canoe.ca.
MacMillan Bloedel Ltd. 1987. *Annual Review*. 4,8.
—. 1990. *Annual Statutory Report*. 15-16.
—. 1991. *Annual Review*. 23.
—. 1998a. "MacMillan Bloedel to Phase Out Clear-cutting; Old-Growth Conservation Is
Key Goal; Customers to Be Offered Certified Products." Vancouver: MacMillan Bloedel.
—. 1998b. "MB Plans to Stop Clear-cutting." *MB Journal* 18(7): 2-6.
—. 1999. "MB Division First to Clear Audit for Forestry Certification Standard." Press
release. Vancouver: MacMillan Bloedel.
MacMillan Bloedel Ltd. and Dovetail Consulting Inc. 1999. "Summary of First Year Critique
Workshop on the MacMillan Bloedel BC Coastal Forest Project." Vancouver: Prepared
by Dovetail Consulting for MacMillan Bloedel.
Matas, Robert, and Patricia Lush. 1998. "How a Forestry Giant Went Green." *Globe and
Mail* 13 July: A1, A6.
Moffat, Andrea C. 1998. "Forest Certification: An Examination of the Compatibility of
the Canadian Standards Association and Forest Stewardship Council Systems in the
Maritime Region." MES thesis, Dalhousie University, Halifax.
"The New Growth at Weyerhaeuser." 1995. *Business Week* 19 June: 63-64.
Noss, Reed. 1989. "Strategies for Conservation of Old Growth." *Forest Planning Canada*
5(4): 8-11.
Oliver, Christine. 1991. "Strategic Responses to Institutional Processes." *Academy of
Management Review* 16(1): 145-79.
Perry, David. 1999. "Proposed Change of Control from MacMillan Bloedel to Weyer-
haeuser: Report to the Minister of Forests on Public Input." BC Government, Victoria.
Raizada, Rachana. 1998. "Corporate Responses to Government and Environmental
Group Actions Designed to Protect the Environment." PhD diss., University of British
Columbia, Vancouver.
Reed, Don. 1999. "Exploring Environmental Communication: Targeting the Investor
Community." In Eric Hansen (ed.), *Environmental Marketing: Opportunities and Strategies
for the Forest Products Industry*. Portland: Oregon State University.
"Roseburg Forest Products in $303 Million Deal." 1995. *New York Times* 29 November.
Salazar, Debra J. 1988. *Comparative Analysis of State Forest Practice Regulations*. Seattle:
College of Forest Resources, University of Washington.
—. 1989. "Regulatory Politics and Environment: State Regulation of Logging Practices."
Research in Law and Economics 12: 95-117.
Smith, Andrew A., Margaret A. Moote, and Cecil R. Schwalbe. 1993. "The Endangered
Species Act at Twenty: An Analytical Survey of Federal Endangered Species Protection,"
Natural Resources Journal 33: 1027-76.
Stanbury, W.T., Ilan B. Vertinsky, and Bill Wilson. 1995. *The Challenge to Canadian Forest
Products in Europe: Managing a Complex Environmental Issue*. Vancouver: Forest
Economics and Policy Analysis Research Unit, University of British Columbia.
Stephens, Tom. 1999. "Speaking Notes from Tom Stephens, President and CEO,
MacMillan Bloedel." MacMillan Bloedel, Vancouver.

Tripp, D., A. Nixon, and R. Dunlop. 1992. *The Application and Effectiveness of the Coastal Fisheries Forestry Guidelines in Selected Cut Blocks on Vancouver Island*. Prepared for the Ministry of Environment, Lands and Parks. Nanaimo, BC: D. Tripp Biological Consultants Ltd.

Vertinsky, Ilan B., and Charlene Zietsma. 1998. *Corporate Greening and Environmental Protection Performance: Static and Dynamic Analysis*. Vancouver: Faculty of Commerce and Business Administration, University of British Columbia. Presented at the Academy of Management Conference, 5-7 August, San Diego, CA.

Vos, Robert O. 1997. "Introduction: Competing Approaches to Sustainability: Dimensions of Controversy." In Sheldon Kamieniecki, George A. Gonzalez, and Robert O. Vos (eds.), *Flashpoints in Environmental Policymaking: Controversies in Achieving Sustainability*. 1-27. Albany: State University of New York Press.

Wallinger, S. 1995. "AF&PA Sustainable Forestry Initiative: A Commitment to the Future." *Journal of Forestry* 48(9): 16-19.

"Weyerhaeuser Builds New Plant." 1995. *Wall Street Journal* 9 March: B6.

"Weyerhaeuser to Cut Production." 1998. *Wall Street Journal* 29 July: 1.

"Weyerhaeuser Signs Project XL Agreement." 1997. *Environmental Solutions* 11 March: 11.

Weyerhaeuser Timberlands USA. 1997. *Vision, Values, and Goals*. Tacoma, WA: Weyerhaeuser Company.

Weyerhaeuser USA. 1995. *1994 Annual Environmental Performance Report*. Tacoma, WA: Weyerhaeuser Company.

—. 1997. *Weyerhaeuser 1996 Annual Environmental Report*. Tacoma, WA: Weyerhaeuser Company.

Wilson, Jeremy. 1990. "Wilderness Politics in B.C.: The Business-Dominated State and the Containment of Environmentalism." In William D. Coleman and Grace Skogstad (eds.), *Policy Communities in Canada: A Structural Approach*. 141-69. Mississauga: Copp Clark Pitman.

—. 1998. *Talk and Log: Wilderness Politics in British Columbia*. Vancouver: UBC Press.

World Commission on Environment and Development. 1987. *Our Common Future*. Oxford: Oxford University Press.

Wright, Don. 1995. "Residuals, Scarcity, Monkeys, and Timber Supply: Some Reflections on Developments in British Columbia."

Yaffee, Steven Lewis. 1994. *The Wisdom of the Spotted Owl: Policy Lessons for a New Century*. Covelo, CA: Island Press.

Zimmerman, Adam H. 1997. *Who's in Charge Here Anyway? Reflections of a Life in Business*. Toronto: Stoddart.

Part 3: Voices

In a war defined by two sides, many actors have been ignored. The chapters in Part 3 focus on actors who have received little or no attention from analysts of forest conflict in the Pacific Northwest and who have been under-engaged by the institutions that have dominated the debate. Authors in this section voice some of the concerns of Native peoples and workers in the emerging nontimber forest economy. These chapters represent a preliminary attempt to broaden the regional conversation about forests and to see how the inclusion of other voices might change that conversation.

Unlike the chapters in the previous section, each of the chapters here focuses on only one side of the Canada-US border. The legal standing of Indigenous people, especially with regard to land, differs substantially between the two countries, and a comparison would miss important details in each jurisdiction. Moreover, the authors of the chapters in this section have been intimately involved in the events that they describe. Thus, the Boyd and Williams-Davidson chapter focuses on British Columbia. Native land claims in the province are so substantial that any effort to reform forest management will necessarily depend on how these claims are resolved. The province is now engaged in land claims negotiations with many First Nations, and their outcomes will have tremendous implications for how forests are managed and used in the future. Thus, all forest stakeholders in British Columbia have paid close attention to these negotiations. Moreover, the first agreement, with the Nisga'a First Nation, has been the subject of considerable controversy both within the Native community and among the general public.

David Boyd and Terri-Lynn Williams-Davidson argue that the realization of Native sovereignty over land claims in British Columbia will promote more sustainable forest management. They also contend that both litigation and negotiation are appropriate and necessary strategies for resolving land claims. Thus, they offer an analysis that is at once legal, political, and cultural. They describe the role of forests in the cultures of Pacific coast Native peoples, review the history of relations between the Canadian state and Indigenous peoples with respect to land, and assess the prospects for fuller realization of Native rights to forest land.

Although the border is likely not so important in defining the situation of nontimber forest workers in the two countries, very little research has been done on this group of people. Neither social scientists nor government statisticians have adequately described the people involved in harvesting forest-floor products. Thus, Beverly Brown analyzes the situation of forest-floor workers only in the Pacific Northwest where she has worked to support organizational efforts of these workers.

Brown's chapter focuses on forest workers involved in nontimber harvesting activities (e.g., tree planting, berry picking, mushroom and herb harvesting). Brown argues that nontraditional forest products are becoming increasingly important in the forest economy. She describes the demography of this workforce as well as working conditions in the forest-floor economy. Finally, she critically evaluates several strategies for integrating the interests of these workers into forest policy making.

By highlighting these overlooked actors, we hope to move toward two goals. The first is to characterize the forest politics more carefully in order to reveal the multiple dimensions of forest conflict. The second is to identify issues of social justice that have been ignored for too long in the forest policy community.

5

Forest People: First Nations Lead the Way toward a Sustainable Future

David R. Boyd and Terri-Lynn Williams-Davidson

> I am trying to save the knowledge that the forests and this planet are alive, to give it back to you who have lost the understanding.[1]

The longest running and most important land conflict in BC history remains unresolved. The Aboriginal people living in this place when Europeans arrived in the eighteenth century never ceded, surrendered, bartered, or signed away their rights to use and occupy the land. Today First Nations are striving to regain control of the land and its resources. Having established Aboriginal rights to fish and wildlife, First Nations are now focusing their efforts on forests. Like fish and wildlife, forests were and are an integral part of Aboriginal society – culturally, economically, spiritually, and aesthetically.

From the perspective of regaining control over forests, there are really two main objectives for First Nations. First, there is a need to end the ecological and cultural destruction caused by industrial logging. This goal can be achieved by moving toward a sustainable ecosystem-based management regime. Second, there is a need to allocate forest resources to First Nations and to distribute the economic benefits of forestry activities more justly, with greater return to local communities. This second goal can be achieved through tenure reform and by changing from provincial ownership to First Nations ownership of the forests.

To regain control over forests, First Nations are using two main strategies: the negotiation of treaties and litigation that seeks recognition of their Aboriginal rights. After over a century of ducking the issue, the provincial and federal governments have agreed to participate in a modern treaty process. However, because of serious problems with the treaty process, some First Nations are using lawsuits to seek recognition of their Aboriginal rights. Both negotiation and litigation have the potential not only to advance the aspirations of First Nations but also to promote a more sustainable approach to managing BC forests.

First Nations and Forests: 10,000 Years of Sustainable Use

The cultures of First Nations[2] in British Columbia are inextricably linked to old-growth forests and have been thus linked for at least 10,000 years.[3] Traditional use of forests by First Nations has been extensive: the existence of the Haida, the Nuu-Chah-Nulth, and other First Nations on the coast has been dependent to a large degree on the wealth of the old-growth forests. Trees were and are used to build houses, make canoes, and carve totem poles. Other objects made from raw materials provided by trees include clothing, baskets, mats, cooking "pots," masks and other ceremonial objects, rope used for fishing and hunting, storage boxes, and tools for cooking, hunting, and fishing.

Old-growth forests are also an essential source of many substances for food and medicine as well as fuel for cooking and heating. Healthy old-growth forests provide critical habitat for wildlife and fish, especially salmon, which are of great importance to First Nations both materially and spiritually. Less tangibly, old-growth forests are also important for spiritual and ceremonial values as places for "vision quests" and education. As described by ethnobotanist Nancy Turner, "Many plants were particularly important culturally, whether it be for food, material, medicine, and/or in playing some role in ceremony or religious thought. However, even the plants that might be considered unimportant by virtue of not being 'used' in some particular way were regarded as special and as living entities with their own power, their own spirit and their own ability to help those deemed to be good and respectful. In fact, everything in the Haida universe – plants, animals, water, rocks, mountains, stars, and the sun and moon – all of these were seen as sacred and important."[4]

First Nations culture continues to be defined by the land, oceans, rivers, and forests. As Turner explains, "The mountains, the waters, the plants and the animals of Haida Gwaii are all part of a magnificent system, supporting and nourishing the Haida and, in turn, respected and embraced by them as an integral part of their culture and identity."[5]

The Nuu-Chah-Nulth of Vancouver Island also reflect the widely held First Nations philosophy that all things are sacred and deserve to be treated with respect. The Nuu-Chah-Nulth phrase *hishuk ish ts'awalk* ("everything is one") embodies the concepts of sacredness and respect:[6] "Nothing is isolated from other aspects of life surrounding it and within it. This concept is the basis for the respect for nature that our people live with, and also contributed to the value system that promoted the need to be thrifty, not to be wasteful, and to be totally conscious of your actual needs in the search for foods. The idea and practices of over-exploitation are deplorable to our people. The practice is outside our realm of values."[7]

The Nuu-Chah-Nulth also have a system of hereditary land stewardship called <u>ha</u>huulhi[8] that integrates ownership, control, and management. Together, *hishuk ish ts'awalk* and <u>ha</u>huulhi form the basis for a stewardship ethic that is most closely embodied in the modern concept of sustainable, community-based ecosystem management.

While the distinctive cultures of First Nations owe much of their history and evolution to the wealth of the forests, their future is also dependent on the continued health of old-growth forests. Without them, the language, culture, and traditional way of life may disappear. Thus, finding a path that leads to sustainability – ecological, cultural, social, and economic – is of the utmost importance to First Nations.

Examining the status of Indigenous languages on the West Coast provides a compelling illustration of the direct connection between healthy traditional Aboriginal cultures and healthy old-growth forests. According to a recent study, "44 out of 68 language groups believed to have been spoken at the time of European exploration are today extinct or spoken by fewer than ten individuals."[9] The extinct languages are mainly in California, Oregon, and Washington, while living languages are found in British Columbia and Alaska. There appears to be a direct correlation between the demise of Indigenous languages and the extent of industrial development in a watershed. The higher the percentage of industrial development, the greater the likelihood that the Indigenous language is extinct.[10]

Industrial Logging in BC Forests: A Century of Destruction

Temperate forests are the most endangered forests on Earth.[11] Coastal temperate rainforests occupy only 0.2 percent of the Earth's surface, yet they contain more biomass than any other place on Earth. British Columbia contains over half of the remaining old-growth temperate rainforest in North America and one-quarter of the world's remaining coastal temperate rainforest.

After millennia of sustainable use by First Nations, there has been an exponential increase in logging in the past 100 years, particularly since 1960. Two-thirds of British Columbia's coastal rainforest – one of the Earth's biologically richest ecosystems – have been degraded by logging or other industrial development. For example, on Vancouver Island, only 11 of 170 watersheds have not been subjected to the damage inflicted by industrial development.[12]

British Columbia manages its forests according to the philosophy of "sustained yield forest management," which is based on liquidating old-growth forests and replacing them with faster growing tree farms. Trees that would naturally live for 500 to 1,000 years are expected to be harvested every sixty to eighty years. The natural range of species will be dramatically modified. On Haida Gwaii (the Queen Charlotte Islands), western red cedar typically makes up about 30 percent of the old-growth forest prior to logging. In the second-growth tree farms, cedar is expected to make up about 2 percent of the crop.[13] The implications for traditional Indigenous cultures are both obvious and devastating – First Nations cannot make a totem pole or a canoe from an eighty-year-old tree, particularly when it is the wrong species for carving.[14]

The unsustainable rate of logging in British Columbia is also a major concern to First Nations. According to the Ministry of Forests, the long-term sustainable rate of logging for the province is between 50 and 60 million cubic metres.[15] The current allowable annual cut is roughly 71 million cubic metres, 20 to 40 percent above the government's calculation of what is sustainable.[16] To make matters worse, most forest ecologists agree that, if ecosystem management were to replace the current industrial-logging paradigm, the sustainable rate would be considerably lower than the ministry's estimate of 50 million cubic metres.[17]

In 1994, the provincial government enacted the Forest Practices Code of British Columbia,[18] intended "to change the way forests are managed" and to make the BC logging industry more sustainable. The code has changed the way that forests are managed, but not to the degree that First Nations and the public had expected. To a large extent, the code was crippled at the outset by a policy directive that its impacts could not result in a 6 percent decrease to the allowable annual cut. Given that the provincial rate of logging is 40 percent above the government's own estimate of the sustainable rate, a 6 percent decrease in the rate of logging falls far short of achieving sustainability.

Audits conducted since the code was enacted show that clear-cutting has continued to be the harvesting method for over 90 percent of cut-blocks,[19] that a large percentage of streams continue to be misclassified or are simply not protected,[20] that clear-cutting continues to predominate on steep, unstable slopes,[21] and that key wildlife protection mechanisms have not been implemented.[22] While there have been some modest improvements in forest practices, there is still a wide gulf between current practices and sustainable ecosystem management.

The demonstrable impacts of industrial logging on BC ecosystems are accumulating. The province has lost at least 142 salmon runs in this century, and another 624 are on the brink of extinction.[23] Logging is a major culprit in the decline of salmon runs. The provincial Ministry of Environment lists 75 endangered or threatened animal species, 241 endangered or threatened plant species, and another 420 vulnerable animal and plant species.[24] Logging is identified by the ministry as the third largest factor in endangering species in the province.[25] As well, the provincial government has estimated that the cost of restoring watersheds damaged by logging will be between $1 billion and $4 billion.[26]

Traditional First Nations territories contain most of the "productive forest land" in British Columbia. Thus, the brunt of the ecological impacts of industrial logging is often borne by First Nations. For example, "forest practices in Clayoquot Sound have contributed to mass wasting of the soil, sedimentation, and reduced fish stocks."[27] First Nations throughout the province, on the coast and in the interior, have suffered great hardships, both economically and culturally, because of declining salmon populations (due in part to logging). Logging and road building also have impacts on wildlife, as described by Ditidaht Hereditary Chief Queesto: "In the early days we used to hunt elk, deer and bear right here by the San Juan River. They were all so plentiful, you could get anything you wanted. I can remember when you would see bands of wolves up along the river ... We always had plenty of game for food. Ever since logging came, there's been no more deer or wolf or elk or beaver. They've all disappeared. Maybe they've been killed off, or maybe they've just moved on to somewhere else. We don't know where the animals have gone."[28]

There is also a decided lack of protection for the cultural heritage of First Nations under the Forest Practices Code. Areas of significance to First Nations routinely continue to be logged, even when the last intact watershed in a First Nation's traditional territory is at stake.[29] As for the participation of First Nations in planning and approval processes, there has been little on-the-ground progress despite several important legal victories.[30] The Ministry of Forests continues simply to notify First Nations of planned logging and road building. In the vast majority of areas being logged, the level of knowledge at the ministry about cultural, spiritual, and ecological values is minimal, yet status quo industrial logging continues.

Government and industry treatment of culturally modified trees (CMTs) illustrates the problem. CMTs are ancient trees that provide physical evidence of First Nations use such as bark stripping, plank removal, and single-tree selection logging for canoes, totem poles, or longhouses. First Nations view CMTs as evidence of their ownership and use of the land, as living testaments to their culture, and as storehouses of traditional knowledge.[31] In contrast, the logging industry and the provincial government view CMTs as a source of aggravation. On many occasions, a logging company has simply notified the First Nation of the existence of a CMT, gotten a permit, cut down the tree, and delivered it to the First Nation![32]

In most of the province, there is no forest land set aside for traditional use by First Nations. Carvers and canoe builders must apply for "special use permits" to receive permission for traditional uses of the forests or to buy logs from logging companies.[33]

While First Nations pay the costs of industrial logging, they receive few of the benefits. British Columbia's current tenure system allocates most of the annual timber harvest to a small number of large corporations. A recent study found that "over 86% of B.C.'s public forest resource is held by 20 corporations, and 23% is controlled by just three companies."[34] As a result, control over logging operations, including planning and forest management, is removed from local and community control.[35]

The existing tenure system makes little provision for allocating forests to First Nations for either traditional or commercial activities. A 1991 Task Force on Native Forestry found that First Nations control less than 1 percent of the provincial allowable annual harvest.[36] The Haida Nation, despite its long history of extensive use of old-growth forests, has never had a timber licence, tree farm licence, or any other form of industrial-logging tenure. Aside from allocation problems, First Nations do not benefit from the current industrial-logging regime.

Most of the timber harvested in British Columbia receives a minimal amount of processing or "value added" before being exported. According to recent statistics, "over 90% of B.C.'s forest product exports are shipped in the form of dimensional lumber, pulp, newsprint and paper."[37] The result of this high-volume, low-value-added commodity approach is that few jobs are sustained at a local level and few economic benefits accrue to First Nations. Haida Gwaii provides a compelling illustration of the lack of local benefits provided by industrial logging. In recent decades, tens of millions of cubic metres of timber have been taken from Haida Gwaii's old-growth forests. This timber has a market value worth billions of dollars, yet the Haida have always faced high unemployment rates[38] because they have no tenure and virtually all of the timber leaves the community without any processing or value being added.[39] Assuming that this scenario is typical throughout British Columbia, it appears that First Nations have borne the costs of industrial logging while reaping few of the economic benefits.

While some First Nations individuals have been employed by or continue to be employed by the logging industry, the centrality of the forests to the cultures of First Nations and the destruction of their traditional ways of life leaves many First Nations opposed to industrial logging. Even for First Nations that engage in logging ventures themselves, there is often a divergence of opinion and strong pressure to practise more responsible stewardship.[40] Many First Nations are angered by the loss of and damage to their traditional territories, and they are frustrated with the fact that economic benefits from forests continue to be removed from local communities.

In the fall of 1999, the frustrations of First Nations people reached new heights. Several First Nations in British Columbia began logging on their traditional territories (or on provincial Crown land, depending on one's perspective) without government permits. The Ministry of Forests responded by issuing stop-work orders under the authority of the Forest Practices Code. Legal proceedings were initiated, and several injunctions were issued by the BC Supreme Court to prevent further logging until the underlying legal issues are resolved.[41] Meanwhile, the two largest First

Nations organizations in the province, the First Nations Summit and the Union of BC Indian Chiefs, called for an international boycott of BC timber products.

Significant changes to the forest management regime are necessary to ensure the protection of Aboriginal culture and Aboriginal rights to the forests. A more desirable approach would be "ecosystem-based management," which emphasizes maintaining ecosystem health and the cultures of First Nations. The National Aboriginal Forestry Association has developed a set of Aboriginal Forest Land Management Guidelines.[42] This proposed management regime (1) stresses the importance of community direction and long-term enhancement of social, spiritual, environmental, and economic values; (2) respects all parts of the forest, including plants, animals, soil, air, water, and all forest users; and (3) respects the diversity of Aboriginal communities as distinct societies with their own languages, cultures, values, and customs. Ecosystem-based management is clearly more compatible with First Nations community-based management and stewardship of the environment than the current regime of industrial forestry.[43]

The European Resettlement of British Columbia

Throughout the world, Indigenous peoples have been displaced from their traditional lands – first by the colonial powers and more recently by the transnational corporations of the industrialized world. This displacement has been justified, and continues to be justified, on the principles of conquest, "discovery," or treaties with First Nations. In Canada, the story of colonization, exploitation, and cultural genocide is no different. However, in one respect, British Columbia is unique.

During the seventeenth, eighteenth, and nineteenth centuries, the British Crown concluded treaties with many, but not all, First Nations across Canada. Some First Nations in British Columbia, the Yukon, the Northwest Territories, and to a lesser extent Quebec and Alberta have either not ceded land through treaties or only recently concluded treaties with the federal and provincial governments.

British Columbia is in an anomalous legal situation with respect to First Nations in Canada because the ownership debate in most of the province remains unresolved. Only small portions of Vancouver Island and northeastern British Columbia have had treaties settled.[44]

Because treaties have not yet been concluded, and will probably not be concluded for the next two decades, First Nations hold unextinguished legal rights over a significant proportion of the land comprising British Columbia.[45] These are legal rights to the land derived from historic occupation of the land by First Nations and from the fact that they lived as self-governing peoples, with their own laws and customs prior to the arrival and settlement by Europeans. This leaves the provincial government and natural resource industries in a precarious position while treaties unfold but presents First Nations in the province with unprecedented opportunities for change.

Decolonizing British Columbia's Forests

There are two paths available to BC First Nations in their quest for greater control over the ownership and management of the province's forests: negotiation and litigation.[46] These paths are often portrayed as mutually exclusive in that the federal and provincial governments have suggested that they will not continue negotiations with a First Nation that is advancing its legal interests through litigation. In fact, this either/or categorization is inaccurate. There are a number of First Nations in the BC treaty process that are negotiating and litigating, or have done so, at the same time. These First Nations include the Haida, Cheslatta, Gitxsan, Wet'suwet'en, Tsay Keh Dene, Gitanyow, Sechelt, Musqueam, Squamish, and Tsleil Waututh.[47]

Litigation and negotiation are intricately intertwined, with new developments in each area having profound implications in the other. For example, the Supreme Court of Canada decision in the *Delgamuukw* case has significantly strengthened the negotiating positions of First Nations in the treaty process. On the other hand, the Nisga'a Final Agreement provides a benchmark to other First Nations of the approximate parameters of a settlement reached through negotiation.[48]

Both negotiation and litigation offer First Nations the opportunity to make significant advances in regaining control over both forests and their cultures. However, each approach is also plagued by major drawbacks, as the following analysis illustrates.

Negotiation: A Sustainable Future through Treaties and Interim Measures Agreements?

First Nations in British Columbia never acquiesced in the colonial or provincial government's usurpation of their land rights. For over a century, efforts by First Nations to have outstanding questions about land ownership in the province answered were rejected. Petitions, protests, and litigation from First Nations were met with obfuscation, delay, royal commissions, white papers, legislation,[49] and discrimination by colonial, federal, and provincial governments. This history is an ongoing source of embarrassment to both British Columbia and Canada. Belatedly, the modern treaty process was kick-started by the 1973 Supreme Court of Canada decision in the *Calder* case brought by the Nisga'a Nation.[50] Twenty-five years of negotiations culminated in 1998 with the Nisga'a Final Agreement. The Nisga'a treaty negotiations are unique in that other modern treaty negotiations with BC First Nations did not begin until 1993.[51]

The Nisga'a Final Agreement

The Nisga'a Final Agreement, signed in August 1998 by leaders of the provincial government, the federal government, and the Nisga'a First Nation, transfers ownership of 2,000 square kilometres from the provincial government to the Nisga'a.[52] This area represents a relatively small fraction of the Nisga'a traditional territory. The treaty also gives the Nisga'a new self-government powers (in areas including language, culture, land use, health, and education) and a financial package worth approximately $300 million.[53] By late 1999, the provincial, federal, and Nisga'a governments had all ratified the treaty.

In the area of forestry, under the treaty the Nisga'a are given new management responsibilities and logging rights on Nisga'a lands. All logging on Nisga'a lands must meet or exceed the rules of the BC Forest Practices Code for Crown land, not the weaker rules for logging on private land.[54] However, the code, as described above, leaves much to be desired from a sustainable forestry perspective even on Crown land. The Nass River region has been abused by decades of overcutting. By the provincial government's own estimate, the rate of logging in this area has been roughly three times the sustainable rate in recent years.[55]

Despite this history of logging abuse, the treaty actually limits the ability of the Nisga'a to reduce the rate of logging to sustainable levels, at least in the short term. For the first five years of the treaty, the rate of logging on Nisga'a lands is held constant at 165,000 cubic metres.[56] For years six through nine, the decline is fixed at gradual steps down to 130,000 cubic metres.[57] These restrictions appear to have been dictated by a provincial government intent on defending the logging industry status quo. Unfortunately, the Nisga'a are effectively precluded from quickly making the necessary transition from the volume-based industrial logging of the past to the value-added, ecosystem-based community forestry that is the future.[58]

The BC Treaty Negotiation Process

In 1993, First Nations, British Columbia, and Canada established the BC Treaty Commission to oversee the modern treaty negotiation process. More than fifty First Nations are now at varying stages of the treaty process.[59] It is obvious, based on the enormous stakes, the complexity of the issues, and the Nisga'a experience, that the treaty negotiation process will take years, if not decades, for most First Nations.

It is generally acknowledged, and courts have repeatedly stated, that negotiation should be preferable to litigation in resolving outstanding questions about Aboriginal title and Aboriginal rights in British Columbia. Litigation is widely regarded as expensive, time-consuming, and unpredictable. Unfortunately, so is the treaty process.

A major shortcoming of the treaty negotiation process at present is the provincial government's policy that, for all completed treaties, the amount of land to be transferred to First Nations will not exceed 5 percent of the provincial land base.[60] When one considers that the traditional territories of First Nations cover close to 100 percent of the province, the extent of the problem becomes obvious. Perhaps some First Nations would prefer comanagement of their traditional territories rather than outright ownership of small percentages of those territories.

In anticipation of the lengthy process and the problems that all parties face in the interim while treaties are being negotiated – disappearing resources for First Nations, economic uncertainty for government and industry – the federal and provincial governments and First Nations agreed to negotiate interim measures agreements (IMAs). These IMAs were intended to alleviate First Nations concerns that there would be no resources left at the end of the negotiating process.[61] Of particular concern are old-growth forests, because the logging industry has already taken so much from traditional territories and because the current rate of logging is so far above what can be sustained in the long run.

A successful example of an IMA dealing with forests is the agreement between the Nuu-Chah-Nulth and the BC government regarding resource management in Clayoquot Sound. The IMA created a joint management regime whereby resource management is supervised by the Central Region Board – a newly created administrative body comprised equally of provincial and First Nations representatives. The IMA also requires the Nuu-Chah-Nulth to enter into a joint venture with MacMillan Bloedel, a major logging company. As a result of the IMA, the Nuu-Chah-Nulth now have a far greater role in forest management and will receive a greater proportion of the benefits of economic activity in the region. The Central Region Board has overseen implementation of the recommendations of the Clayoquot Sound Scientific Panel, resulting in significant progress toward sustainable ecosystem management of the globally renowned Clayoquot Sound area.[62]

Unfortunately, the government of British Columbia has been reluctant to negotiate IMAs with many First Nations. This reluctance has been the subject of considerable criticism, and even the arm's-length BC Treaty Commission has publicly condemned the provincial government for effectively undermining the process by refusing to negotiate IMAs.[63] Despite the BC Treaty Commission's moral suasion, the provincial government remains unrepentant in refusing to negotiate IMAs until the latter stages of the treaty process.[64]

Both treaties and IMAs offer major opportunities to First Nations in British Columbia seeking a sustainable future through greater control over forest management. The potential for reforming forest management, forest tenures, and forest practices through treaties and IMAs to reflect the respect for nature inherent in traditional First Nations concepts of stewardship has been demonstrated, to varying degrees, by the Nisga'a Final Agreement and the Nuu-Chah-Nulth Interim Measures Agreement.

Optimism over the potential for change through negotiation must be tempered by several key factors, including the provincial government's reluctance to negotiate IMAs in good faith and in a timely manner, the length of the treaty negotiation process, and the provincial policy limiting to 5 percent the amount of provincial land available for treaty settlements. Unless these critical constraints can be addressed, litigation will continue to be an attractive option to many First Nations in British Columbia, particularly in light of the recent Supreme Court of Canada ruling in the *Delgamuukw* case.[65]

Litigation: A Sustainable Future through the Courts?

Like negotiation, litigation provides a vehicle through which First Nations can acquire greater control over the ownership and management of forests in British Columbia. Perhaps the greatest difference between negotiation and litigation is the element of risk associated with the latter. Litigation is very unpredictable and rigid, resulting in win or lose results. Uncertainty is exacerbated by the fact that the area of Aboriginal law is probably evolving more quickly than any other area of Canadian jurisprudence.[66]

However, two basic facts make litigation an attractive option for First Nations in British Columbia. First, because treaties were never negotiated, First Nations have an unextinguished legal interest in most of the lands and waters of British Columbia. Second, Aboriginal rights were entrenched in Canada's Constitution in 1982,[67] giving First Nations unprecedented legal strength in asserting and defending their rights. The Supreme Court of Canada has repeatedly held that Aboriginal rights are included in Section 35 in order to protect and reconcile the interests of First Nations with Canadian sovereignty.[68]

There is a variety of legal tools available for First Nations seeking to accomplish their cultural, ecological, and economic objectives in reforming the logging industry and achieving sustainability. The most powerful cases involve invoking Aboriginal title, which means asserting collective ownership of the land and the forests based on historic occupation and use. The other category of cases involves invoking Aboriginal rights, which are generally related to specific activities practised by a First Nation such as hunting, fishing, or medicine gathering. In the context of old-growth forests, Aboriginal rights can be claimed as a means of stopping industrial logging that will infringe upon those rights or, more proactively, as a means of using trees and other elements of old-growth forests. It may also be possible to assert a commercial Aboriginal right to old-growth forests where First Nations can establish that trade in canoes, boxes, or other wood products was an integral part of their cultures. Finally, the forest management regime may be scrutinized in court for its infringement of Aboriginal rights and title. Each of these legal tools is discussed in turn.

Aboriginal Title

For over a century, the BC government has managed the logging industry based on the assumption that the government owns all the forests on public land. That fundamental assumption is now acknowledged to be incorrect. Recent court decisions have revealed that ownership of the forests of British Columbia is, to a large degree, unresolved because treaties were never negotiated.

The most important Aboriginal law decision in Canadian history is the *Delgamuukw* decision handed down by the Supreme Court of Canada on 11 December 1997. For the first time, the Supreme Court addressed directly the meaning of Aboriginal title. The decision clarified that Aboriginal title is a legal interest in the land itself, including the minerals beneath the land and the "fruits of the land," such as forests. *Delgamuukw* also ended long-standing speculation that the Aboriginal title held by BC First Nations had been extinguished by colonization or settlement.

According to the Supreme Court of Canada, Aboriginal title is largely comparable to ordinary property ownership. However, there are several important differences:

- Land held pursuant to Aboriginal title is communally owned, not individually owned.
- Land held pursuant to Aboriginal title can only be sold to the federal government.
- Land held pursuant to Aboriginal title is subject to an "inherent limit," meaning that such land cannot be used in a way that impairs its utility for traditional use by future generations.

The Supreme Court's characterization of Aboriginal title is based, to a large extent, on First Nations laws of land stewardship. First Nations hold land collectively, not individually. It is impossible for First Nations to "own" the land, forests, rivers, oceans, lakes, and creatures of these domains ("traditional territories"), since each has a special spirit and its own persona. First Nations "ownership" of traditional territories is more accurately described as stewardship or a hereditary responsibility to manage traditional territories in a manner that ensures availability for future generations.[69]

From a forestry perspective, one of the most interesting aspects of the Supreme Court's decision in *Delgamuukw* is the principle that land-use activities on lands under Aboriginal title are subject to an "inherent limit,"[70] which means that lands cannot be used in a manner that destroys the special bond between Aboriginal people and those lands. The court used strip mining in a traditional hunting area and paving over a burial site as examples of destructive activities that would be prohibited by the unique nature of Aboriginal title.

The Supreme Court bases the concept of an inherent limit on the "special relationship" between First Nations and the land and resources (the fact that First Nations cultures are inextricably intertwined with the environment).[71] The rationale for incorporating inherent limits into Aboriginal title is the same as that which underlies the concept of sustainability – namely, acting in a manner that does not compromise the environment to the disadvantage of future generations. In the words of the Supreme Court, "The law of Aboriginal title does not only seek to determine the historic rights of Aboriginal peoples to land; it also seeks to afford legal protection to prior occupation in the present-day. Implicit in the protection of historic patterns of occupation is a recognition of the importance of the continuity of the relationship of an Aboriginal community to its land over time."[72]

The Supreme Court appears to be suggesting that only sustainable activities will be allowed on Aboriginal lands. Furthermore, if First Nations are bound by this inherent limit, then it follows that other parties operating on Aboriginal lands must be similarly constrained. This regulation raises many questions, including whether clear-cutting of old-growth forests is a sustainable activity.

While the concept of an inherent limit is applauded, it should not be confined only to lands under Aboriginal title. The history of dominant society's industrial logging, resource use, and relationship with the land and the environment is remarkably different from the traditional relationship that First Nations have with the land and its resources. However, we are all dependent on the land and the environment. Regardless of race or culture, we all depend on clean air to breathe, clean water to drink, and food grown on the land. Accordingly, all of our activities should be governed by the common-sense principle of inherent limits.

The Supreme Court ruled that courts must give equal weight to oral evidence in cases involving First Nations.[73] The court also provided considerable guidance on the evidence required to prove Aboriginal title. The land in question must have been occupied prior to the British assertion of sovereignty in 1846. There must be continuity of occupation, though not necessarily unbroken continuity.[74] Occupation of the land in question must have been exclusive (although the Supreme Court acknowledged the possibility of "shared exclusivity").[75]

In *Delgamuukw*, the Supreme Court also ruled that there is an "inescapable economic aspect" to Aboriginal title.[76] This ruling is relevant to forestry in that the government has a duty to ensure that First Nations are given an opportunity to share in the economic benefits.[77]

Yet another fundamental principle enshrined in the *Delgamuukw* decision is that both the federal and the provincial governments owe a fiduciary duty to First Nations.[78] This duty means that governments have a trust-like obligation to handle dealings with First Nations honourably. In the forestry context, the practical implication of the provincial government's fiduciary duty is that the government must consult with a First Nation prior to approving logging activity in its traditional territory. At a minimum, the Supreme Court said, the consultation must be meaningful and should substantially address the concerns of the First Nation. When Aboriginal rights are threatened, consent of the First Nation is required;[79] the Supreme Court held that an important aspect of Aboriginal title is the right to choose how lands are used. Although the precise parameters of what constitutes meaningful consultation remain to be determined, the provincial government's duty to consult clearly gives First Nations legitimate opportunities to incorporate their ecological and cultural concerns into the forest management process.

As outlined above, the existing forest tenure system in British Columbia effectively excludes First Nations. Courts are beginning to recognize that the exclusion of First Nations is problematic, as shown by the BC Court of Appeal in *MacMillan Bloedel* v. *Mullin*: "There is a problem about tenure that has not been attended to in the past. We are being asked to ignore the problem as others have ignored it. I am not willing to do that."[80]

The exclusion of First Nations from the forest tenure system is difficult to reconcile with the fact that even early Canadian Aboriginal law recognized that an Aboriginal interest in land includes the forests and other resources on that land. One of the earliest decisions addressing Aboriginal rights was *St. Catherines Milling and Lumber Co.* v. *The Queen* in 1888.[81] In its decision, the court held that, upon surrender of Aboriginal rights and title in the treaty process, the rights to the timber passed from the Aboriginal people to the provincial government, not the federal government. It follows logically, then, that in the absence of a treaty the rights to the timber remain with the Aboriginal people. This hypothesis was supported by the *Calder* decision,[82] in which the Supreme Court of Canada held that Aboriginal title includes the "right to occupy the lands and to enjoy the fruits of the soil, of the forest and of the rivers and streams."

A recent BC Court of Appeal decision in a case brought by the Haida Nation provides further evidence that an Aboriginal interest in land includes the trees on that land.[83] The court ruled that Aboriginal title is a valid legal encumbrance on provincial title to the land and the forests. This decision casts doubt on the legal validity of exclusive logging licences, such as tree farm licences in British Columbia. An exclusive logging licence simply cannot be reconciled with Aboriginal ownership of the forests.[84] This decision, along with *Delgamuukw*, casts a shadow of legal uncertainty over the entire forestry management regime in Canada.

Attempting to prove Aboriginal title through litigation is expensive and daunting though not dissimilar from the effort required to negotiate a treaty. The Gitxsan and Wet'suwet'en First Nations began preparing for the *Delgamuukw* case in the mid-1970s. The trial did not begin until the late 1980s. The 1997 *Delgamuukw* decision did not actually decide the extent of Aboriginal title held by the Gitxsan and Wet'suwet'en First Nations because the Supreme Court held that the trial judge had failed to consider adequately the oral evidence brought forward by the First Nations. The Supreme Court decision sent the two First Nations back to the beginning (trial court), albeit with a much stronger legal and evidentiary framework.[85] Chief Justice Lamer added these final words: "Ultimately it is through negotiated settlements, with good faith and give and take on both sides, reinforced by judgments of this Court, that we will achieve ... the reconciliation of the pre-existence of Aboriginal societies with the sovereignty of the Crown. Let us face it, we are all here to stay."[86]

Nevertheless, for First Nations unhappy with the constraints of the treaty negotiation process, and particularly the provincial government's policy that only 5 percent of the land in the province will be returned to First Nations, going to court to assert Aboriginal title remains an attractive option. Given the principle of inherent limits, the more land found to be subject to Aboriginal title the better, from cultural and ecological perspectives.

Aboriginal Rights to Forests

For the purposes of this discussion, Aboriginal rights will be limited to rights to engage in particular activities as distinct from Aboriginal title, which is a legal interest in land.[87] Many of the Aboriginal rights cases in Canada have been concerned with hunting and fishing. Only recently have cases involving Aboriginal rights to the forests begun to emerge.[88] These cases are still embroiled at varying stages in the legal system and will not be discussed in detail here because as yet there are no definitive judicial pronouncements on this topic. The Supreme Court of Canada has never directly addressed the issue of an Aboriginal right to forests. However, it is reasonable to assume that the general principles for proving Aboriginal rights apply to establishing Aboriginal rights to the forests.

There are two distinct categories of potential Aboriginal rights relevant in the context of old-growth forests. The first category involves an Aboriginal right to forests for either traditional or commercial uses. The second category involves Aboriginal rights to a broader range of activities dependent on old-growth forests, from food and medicine gathering to ceremonial and spiritual activities such as vision quests.

Neither a traditional Aboriginal right to forests nor a First Nations commercial right to forests has been established. However, an Aboriginal right to forests or to the use of old-growth forests for a variety of traditional purposes would be based on the same principles as the Aboriginal right to fish. The Supreme Court of Canada ruled, in *Sparrow* and subsequent cases, that First Nations have an Aboriginal right to fish that is constitutionally protected.[89] The court based these decisions on evidence that fishing was and is a practice or custom integral to the distinctive culture of the particular First Nation claiming the right. The court went on to say that conservation is the top priority in fisheries management. Then, in allocating fish among various groups, Aboriginal people are to be given priority over other users because of the historic reliance of First Nations on fish, the integral role of fish in their cultures, and the constitutional status of their rights. Governments are allowed to infringe upon Aboriginal rights, but they require a valid legislative objective and must act in a manner consistent with their fiduciary duty to Aboriginal people.

The Supreme Court recently held, in the *Van der Peet* and *Gladstone* decisions, that commercial rights to use natural resources may be held by First Nations in certain circumstances. The legal test for proving a commercial Aboriginal right is the same as for proving a traditional Aboriginal right. A First Nation must prove that a practice, tradition, or custom is "integral" to its "distinctive culture."

Proving a commercial Aboriginal right to forests will not be easy. Despite extensive evidence that the Sto:lo First Nation engaged in the harvesting and trading of salmon, the Supreme Court held in *Van der Peet* that a commercial right to fish had not been established.[90] However, in *Gladstone* a Heiltsuk fisherman was able to establish successfully a commercial Aboriginal right to fish by proving that trading a seafood product known as herring spawn on kelp was an integral part of Heiltsuk culture.[91] The Supreme Court distinguished these two cases by holding that the Heiltsuk harvested herring spawn on kelp not only to meet their own needs but also to trade with other First Nations. This "excess" was an integral part of the Heiltsuk culture.

A recent decision in the controversial *Marshall* case interprets a 1760 treaty as granting the Mi'kmaq First Nation the right to earn a "moderate livelihood" from hunting, fishing, and gathering.[92] The Supreme Court of Canada overturned the conviction of Donald Marshall for selling eels, based on his treaty right to earn a living. The *Marshall* decision has limited application in British Columbia because it is based on a specific treaty, but it will no doubt be used to buttress assertions of commercial Aboriginal rights.

Among the First Nations of coastal British Columbia is abundant anthropological evidence of extensive trading of canoes, masks, headdresses, and wooden boxes. For example, noted Haida artists Robert Davidson and Bill Reid have called Massett Inlet on Haida Gwaii the "bent-box capital of the world." The Haida were renowned for the quality of their canoes, which were traded up and down the coast.[93] Clearly, there is potential for establishing not only a traditional Aboriginal right to the forests but also a commercial Aboriginal right to the forests.[94]

Reforming Industrial Logging

Both Aboriginal title and Aboriginal rights can be used to reform industrial logging and concentrated corporate ownership. Lawsuits have been brought by First Nations to stop the logging of a specific area;[95] the granting, transfer, or renewal of a licence;[96] or the approval of a logging plan.[97] Such lawsuits rely on evidence that either Aboriginal title or Aboriginal rights, which are constitutionally protected, would be violated. As the BC Court of Appeal said in *MacMillan Bloedel* v. *Mullin*, "The proposal is to clear-cut the area. Almost nothing will be left. I cannot think of any native right that could be exercised on lands that have recently been logged."[98] Such infringements must be minimized by the government[99] and must be based on a compelling objective, such as economic development.

Three recent Supreme Court of Canada decisions (*Gladstone, Adams,* and *Marshall*) open the door to scrutinizing not only selected sections of resource management legislation but also the entire legislative and regulatory system. In each case, the court suggests that resource management regimes that fail to make provisions for Aboriginal rights could be struck down.[100] In *Marshall*, the court stated that "Specific criteria must be established for the exercise by the Minister of his or her discretion to grant or refuse licences in a manner that recognizes and accommodates the existence of an Aboriginal or treaty right."[101] In *Adams*, in a passage repeated in *Marshall*, the court stated that

> Parliament may not simply adopt an unstructured discretionary administrative regime which risks infringing Aboriginal rights. If a statute confers an administrative discretion which may carry significant consequences for the exercise of an Aboriginal right, the statute or its delegate regulations must outline specific criteria for the granting or refusal of that discretion which seek to accommodate the existence of Aboriginal rights. In the absence of such specific guidance, the statute will fail to provide representatives of the Crown with sufficient directives to fulfil their fiduciary duties, and the statute will be found to represent an infringement of Aboriginal rights under the *Sparrow* test.[102]

The BC Forest Act[103] makes no provision for First Nations tenure or licences.[104] The BC Forest Practices Code,[105] as discussed above, provides little or no protection for Aboriginal rights or First Nations culture. There are no explicit criteria to guide decision makers in avoiding the infringement of Aboriginal rights. As a result, the *Gladstone*, *Adams*, and *Marshall* cases support broad challenges to the entire BC forest management regime.

It is remarkable that, in both Aboriginal title and Aboriginal rights cases (notably *Delgamuukw* and *Sparrow*), the Supreme Court of Canada has emphasized legal principles that prioritize sustainability and conservation. In *Delgamuukw*, the court placed an inherent limit on land held under Aboriginal title – meaning that land cannot be used in a way that interferes with the opportunities of future generations to use the land in traditional ways.[106] In *Sparrow*, the court ruled that all fisheries management decisions must place conservation ahead of any allocation of fish among would-be fishers. In *Marshall*, the court reiterated that in fisheries management "the paramount regulatory objective is conservation."[107]

This conservation-first approach applies not only to Aboriginal fisheries management but also to all Canadian fisheries management decisions.[108] Thus, the federal government's controversial ban on coho fishing in the summer of 1998 was justified by federal fisheries minister David Anderson on the basis that he had a legal (and ethical) duty to put conservation first.[109] In 1999, even broader salmon-fishing closures were ordered by the federal Department of Fisheries and Oceans in order to comply with the conservation-first mandate.

The implications of these two decisions in the context of First Nations concerns about the future of British Columbia's old-growth forests are extremely promising. One can argue persuasively that lands under Aboriginal title cannot be subjected to industrial logging because the ecological and cultural impacts would violate Aboriginal title, Aboriginal rights to the forests, and the Supreme Court's principle of inherent limits. In contrast, community-based ecosystem management would probably meet the inherent limit test. If the reasoning in *Sparrow* regarding fisheries management could be adapted to forest management, then there would be profound repercussions for industrial logging. Major changes would be necessary in order to put conservation first in forest management. For example, decisions setting the annual rate of logging above the sustainable rate or imposing a 6 percent limit on the impacts of the Forest Practices Code would likely be vulnerable to legal challenges. Similarly, decisions allowing logging to occur every sixty to eighty years and clearcutting to occur in areas where cedar or other species important to First Nations cultures do not regenerate would be vulnerable to legal challenges.

It is intriguing that these conservation-first legal doctrines have evolved in the realm of Aboriginal law. This evolution toward law informed by First Nations worldviews and ecology may be a tremendous but largely unrecognized contribution to Canadian law from First Nations law and culture. It would be ironic yet just if the judicial system, widely regarded as deeply conservative and antithetical to Aboriginal values, were to play a leadership role in integrating First Nations law and philosophy about stewardship into the Canadian legal and political systems.[110] While the stewardship ethic of First Nations is sometimes criticized as romantic and idealistic, the Supreme Court of Canada is clearly taking it at face value.

Conclusion

First Nations in British Columbia are forest people – their cultures are firmly rooted in the old-growth forests. For thousands of years, the First Nations in the province lived in harmony with the region's forests, depending on the wealth of the old-growth forests for material, cultural, and spiritual needs. Over the past century, that healthy relationship has been shattered by the intrusion of industrial logging – a profit-driven juggernaut that has had catastrophic ecological and cultural consequences. To paraphrase the Supreme Court of Canada, reconciling the rights of First Nations with Canadian society means that forests, which make First Nations what they truly are, cannot continue to be destroyed, thereby destroying First Nations cultures.

At the beginning of the twenty-first century, the pendulum is swinging back toward harmony or, in more modern parlance, sustainability. Traditional First Nations wisdom about stewardship is poised to play a greater role in BC forest management as change sweeps through the unsustainable logging industry. Whether through negotiation or litigation, First Nations have a tremendous opportunity to regain control over the ownership and management of forests in British Columbia.

Recognition of Aboriginal title, Aboriginal rights to forests, and traditional ecological knowledge will stimulate reform of the forest tenure system and a move away from industrial logging toward ecosystem-based community forestry. The transition will be controversial, difficult, and fiercely opposed by those with vested interests in the status quo. The changes are necessary, however, to maintain ecological and cultural integrity in British Columbia. In the long run, First Nations, local communities, forest workers, the forests, and the diversity of life will benefit as forestry becomes a sustainable activity. The new sustainable forestry will be guided by First Nations traditional philosophies of stewardship, such as *hishuk ish ts'awalk* and *hahuulhi*.

Notes

1 Paiakan, Kayapo Indian, Brazil, as cited in Julian Burger (ed.), *Gaia Atlas of First Peoples* (London: Robertson McCarta, 1990), 32.
2 We have chosen the term "First Nation" to refer to the Aboriginal people of British Columbia and Canada. It includes Indian, Inuit, and Métis peoples.

3 For example, oral history of the Haida documents when Haida people inhabited Haida Gwaii. Certain clans of the Haida Nation have the right to wear a tree as a crest since they witnessed the first tree to grow on Haida Gwaii. Archeological and scientific evidence establishes the first appearance of trees and forests about 10,000 years ago. See G. Pellatt and R.W. Mathewes, "Paleoecology of Postglacial Tree Line Fluctuations on the Queen Charlotte Islands, Canada," *Ecoscience* 1,1 (1994): 71-81.

4 N. Turner, "Plants of Haida Gwaii," 27 [unpublished typescript].

5 Ibid., 11.

6 Clayoquot Sound Scientific Panel, "First Nations' Perspectives Relating to Forest Practices Standards in Clayoquot Sound," March 1995, 6 [unpublished typescript].

7 Roy Haiyupis, "Ecosystem Sustainability: A Nuu-Chah-Nulth Perspective," 1 [unpublished typescript, 1994].

8 Clayoquot Sound Scientific Panel, 8-9.

9 Ecotrust, Pacific GIS and Conservation International, *The Rainforests of Home: An Atlas of People and Place, Part 1: Natural Forests and Native Languages of the Coastal Temperate Rainforests* (Portland, OR: Ecotrust, 1995), 8.

10 Ibid. Where over 25 percent of a watershed has been developed, there is a high likelihood that the local Indigenous language has become extinct.

11 Bryant, Dirk, Daniel Nielson, and Laura Tangley, "The Last Frontier Forests: Ecosystems and Economies on the Edge," Washington: World Resources Institute, 1997.

12 Ibid.; and Ecotrust, *Rainforests*.

13 See Council of the Haida Nation, "Special Edition: Forestry on Haida Gwaii," *Haida Laas: Journal of the Haida Nation* (1994).

14 Haida people and other First Nations of the Northwest Coast require cedar trees in excess of 500 years old for the construction of totem poles and canoes. Old-growth trees have a tighter grain and are more suitable for carving. Other ceremonial objects require straight-grained red cedar, yellow cedar, yew, or alder.

15 BC Ministry of Forests, "AAC Listing by Timber Supply Area," 1998; see www.for.gov.bc. ca.

16 Ibid.

17 For example, the Clayoquot Sound Scientific Panel recommended a 70 percent reduction in the rate of harvest, and recent AAC decisions by the chief forester have indeed reduced the AAC in the Clayoquot Sound area by 62 percent. M. Patricia Marchak et al., *Falldown: Forest Policy in British Columbia* (Vancouver: David Suzuki Foundation, 1999), 49.

18 RSBC 1994, c. 40 [hereafter the "code"].

19 See Sierra Legal Defence Fund, *The Clearcut Code* (Vancouver: Sierra Legal Defence Fund, 1996).

20 See Sierra Legal Defence Fund, *Stream Protection under the Code: The Destruction Continues* (Vancouver: Sierra Legal Defence Fund, 1997).

21 See Sierra Legal Defence Fund, *Going Downhill Fast: Landslides and the Forest Practices Code* (Vancouver: Sierra Legal Defence Fund, 1997).

22 See Sierra Legal Defence Fund, *Wildlife at Risk* (Vancouver: Sierra Legal Defence Fund, 1997).

23 Hyatt Slaney et al., "Status of Anadromous Salmon and Trout in British Columbia and Yukon," *American Fisheries Society* October 1996: 20-32.

24 Ministry of Environment, Lands and Parks, *Environmental Trends in British Columbia, 2000* (Victoria: Ministry of Environment, Lands and Parks, 2000).

25 Ibid.

26 P.A. Slaney and A.D. Martin, "The Watershed Restoration Program of British Columbia: Accelerating Natural Recovery Processes," *Water Quality Research Journal of Canada* 33,2 (1997): 325-46.

27 Clayoquot Sound Scientific Panel, "First Nations' Perspectives," 7.

28 Chief Charlie Jones, with Stephen Bosustow, *Queesto: Pacheenaht Chief by Birthright* (Nanaimo: Theytus Books, 1981), 37-38.

29 *Siska Indian Band* v. *B.C. (Minister of Forests)* [1998] 62 B.C.L.R (3d) 133 (S.C.C.); and *Siska Indian Band* v. *B.C. (Ministry of Forests)*, Vancouver Registry No. A992665, unreported decision, 22 October 1999 (S.C.C.).

30 See the decisions discussed later in this chapter: *Delgamuukw* v. *British Columbia* [1997] 3 S.C.R. 1010; *Haida Nation et al.* v. *British Columbia (Ministry of Forests)* [1998] 1 C.N.L.R. 98 (BCCA); and *Halfway River First Nation* v. *B.C. (Ministry of Forests)* [1999] B.C.C.A. 470, unreported decision, 12 August 1999, affirming [1997] 39 B.C.L.R. (3d) 227 (S.C.C.).

31 CMTs are frequently the only source of traditional knowledge for the selection of appropriate trees for canoes and totem poles. Many First Nations elders were removed from their communities and traditional territories at an early age to live at residential schools for years at a time. The missionaries urged First Nations to cut or burn totem poles, and many were stolen or bought by museums and collectors around the world. The forces of residential schools, Christianity, assimilation, and relocation displaced the traditional knowledge of building canoes and totem poles. Only in the past thirty years have First Nations experienced a "renaissance" in canoes and totem poles.

32 Logging companies must obtain permits from the provincial Archaeology Branch to cut CMTs or other heritage sites or objects. These permits have frequently been issued to allow logging companies to engage in road building or logging while avoiding the destruction of CMTs when possible. Otherwise, they may be cut down and dated (CMTs and cultural heritage are only protected under provincial legislation if they are pre-1846). This practice may change with recent decisions such as the *Delgamuukw* decision (discussed later in this chapter) and the *Kitkatla* decision. See *Kitkatla Band* v. *The Ministry of Small Business, Tourism and Culture, the A.G.B.C., and International Forest Products Ltd.* [21 October 1998] Victoria 982223 (B.C.S.C.).

33 These special-use permits are at the sole discretion of district managers of the Ministry of Forests. In many instances, First Nations have been denied special-use permits for traditional uses because of district managers' incredibly narrow interpretation of "traditional uses." However, see the discussion below in the "Aboriginal Rights to the Forests" section regarding challenges to legislative and regulatory systems.

34 C. Burda, D. Curran, F. Gale, and M. M'Gonigle, "Forests in Trust: Reforming British Columbia's Forest Tenure System for Ecosystem and Community Health," University of Victoria Eco-Research Chair in Environmental Law and Policy, 1997, 2.

35 This "isolation from consequences" is discussed in K. Drushka, *Stumped: The Forest Industry in Transition* (Vancouver: Douglas and McIntyre, 1985).

36 Task Force on Native Forestry, "Native Forestry in B.C.: A New Approach," Victoria, 1991.

37 Burda, Curran, Gale, and M'Gonigle, "Forests in Trust," 2.

38 The Haida villages of Skidegate and Massett face average unemployment rates of 80 percent. High unemployment rates are common throughout First Nations communities in Canada, which face rates 25 to 55 percent higher than the rest of the population of Canada.

39 British Columbia Wild, *Taking It All Away: Communities on Haida Gwaii Say Enough Is Enough* (Vancouver: British Columbia Wild, 1996).

40 For a discussion of joint ventures between First Nations and forest companies, see D. Curran and M. M'Gonigle, *First Nations' Forests: Community Management as Opportunity and Imperative* (Victoria: Faculty of Law and Environmental Studies Programme, University of Victoria, 1997). The joint ventures and initiatives taken by First Nations to incorporate ecosystem-based management and nontimber values are usually rejected by the Ministry of Forests, essentially because the proposed volumes and methods reduce the operable land base and do not ensure adequate fibre flow.

41 *R.* v. *Chief Dan Wilson et al.* [1999] Vernon Registry No. 23911, unreported decision, J. Sigurdson, 12 November 1999 (S.C.C.); *R.* v. *Chief Ronnie Jules et al.* [1999] Vernon Registry No. 23911, unreported decision, J. Sigurdson, 12 November 1999 (S.C.C.); and *R.* v. *Chief Ron Derrickson et al.* [1999] Kelowna Registry No. 46440, unreported decision, Sigurdson, J., 12 November 1999 (S.C.).

42 See H. Bombay, *Aboriginal Forest-Based Ecological Knowledge in Canada* (Ottawa: Anishinabe Printing, 1996), and H. Bombay, *An Aboriginal Criterion for Sustainable Forest Management* (Ottawa: Anishinabe Printing, 1995).

43 Professor Frank Cassidy argues that "Sustainable development in British Columbia and Canada as a whole will not be achievable without the full involvement and support of

indigenous peoples. Indigenous peoples are not just one more stakeholder in the process of achieving sustainable development. They have unique collective rights which make them a central part of this process. In addition, they have much knowledge and wisdom to offer. Until the rights, practices, institutions and knowledge of indigenous peoples are fully respected, the goal of sustainable development will continue to be illusive and unachievable. The sooner this fact is recognized, the better." Frank Cassidy, "Indigenous Peoples and Sustainable Development," Centre for Sustainable Regional Development, University of Victoria, 1994, 4. See also the Brundtland Commission, which stated that "Tribal and indigenous peoples will need special attention as the forces of economic development disrupt their traditional lifestyles – lifestyles that can offer modern societies many lessons in the management of resources of complex forest, mountain and dryland ecosystems. Some are threatened with virtual extinction by insensitive development over which they have no control. Their traditional rights should be recognized and they should be given a decisive voice in formulating policies about resource development in their areas." World Commission on Environment and Development, *Our Common Future* (Oxford: Oxford University Press, 1987), 115.

44 Fourteen treaties were concluded with the First Nations on Vancouver Island in the 1850s, constituting about 3 percent of the island's area. In 1999, Treaty 8 was extended into the northeastern portion of British Columbia. For a general history of British Columbia and the struggle for legal and political recognition of Aboriginal title in the province, see P. Tennant, *Aboriginal Peoples and Politics* (Vancouver: UBC Press, 1990).

45 The Nisga'a Treaty took twenty-five years to negotiate; see the discussion below under the section "Negotiation: Toward a Sustainable Future through Treaties and Interim Measures Agreements?"

46 Blockades are also a path frequently taken by First Nations when all other avenues fail. In British Columbia, the extensive use of protest blockades in the 1980s led to the creation of the BC Claims Task Force, the BC Treaty Commission, and the BC Treaty Process. Until then, only the federal government negotiated with First Nations under a lengthy "comprehensive claims policy."

47 BC Treaty Commission, "1998 Annual Report," 1998. See www.bctreaty.net.

48 First Nations in British Columbia support the decision of the Nisga'a Nation from a position of respect for Nisga'a self-determination. However, the Nisga'a Treaty may not be suitable for other First Nations with different histories and needs. If provincial and federal governments are unwilling to negotiate beyond the parameters of the Nisga'a Treaty, then litigation may be the preferred approach for other First Nations.

49 In 1884, the Indian Act was amended to prohibit First Nations from holding potlatch and Sun Dance ceremonies. Potlatches define the cultural, social, and political fabric of First Nations societies in the northwest coast of British Columbia. The Indian Act was also amended in 1927 to prohibit First Nations from hiring lawyers to advance Aboriginal title and rights cases without government approval. Both of these amendments were repealed in 1951. In addition, First Nations did not obtain the provincial franchise until 1947 and the federal franchise until 1960. See, generally, Tennant, *Aboriginal Peoples.*

50 *Calder* v. *Attorney-General of British Columbia* [1973] S.C.R. 313. The Supreme Court of Canada unanimously held that Aboriginal title was a pre-existing legal right but was split as to whether Aboriginal title continued to exist in British Columbia.

51 The BC Treaty Process was designed by the BC Claims Task Force in 1990-91. The BC Treaty Commission formally began to accept statements of intentions to negotiate treaties in 1993.

52 See Nisga'a Final Agreement at www.aaf.gov.bc.ca/aaf/treaty/nisgaa.

53 Ibid.

54 Ibid., Chapter 5.

55 The annual allowable cut for the Nass Timber Supply Area is 1,150,000 cubic metres, yet the long-term harvest level is 410,000 cubic metres per year. See BC Ministry of Forests, "AAC Listing."

56 Nisga'a Final Agreement, Chapter 5.

57 Ibid.

58 From the perspective of sustainability, this is a significant flaw in the treaty. Opinions on the overall merits of the Nisga'a Final Agreement are widely divergent. To some, it represents a fraction of what the Nisga'a really deserve given their thousands of years in the Nass Valley and the indignities inflicted by European settlement. To others, the treaty is far too generous, giving the Nisga'a too much land, too much money, and too much power. While some view it as a template for future treaties, others argue that each negotiation will be different.

59 Not all First Nations in British Columbia are engaged in the treaty process. Those not in the process advocate a different process in which the province is not at the table (since only sovereign nations should negotiate treaties), have become frustrated with the unwillingness to negotiate interim measures agreements (see below) at an early stage in the process, or are engaged in litigation and have had negotiations suspended by either the provincial or the federal government.

60 Ministry of Aboriginal Affairs, "Provincial Approach to Treaty Negotiation of Lands and Resources." This position is purportedly proportionate to the First Nations population of British Columbia but is not proportionate to the amount of land currently used by First Nations or that used in 1846 (the Crown's assertion of sovereignty over British Columbia).

61 The BC Claims Task Force recommended that IMAs were not only an important indicator of sincerity and commitment of the parties to negotiate a treaty but also would protect interests prior to the beginning of negotiations. The task force recommended that parties be able to initiate IMAs at any time in the treaty process. BC Claims Task Force, "The Report of the British Columbia Claims Task Force," 28 June 1991 [unpublished typescript].

62 See note 17.

63 BC Treaty Commission, "Annual Report," 1997; see www.bctreaty.net.

64 The provincial government's position is that IMAs will not be negotiated until stage four of the process, when an agreement in principle is negotiated. Ministry of Aboriginal Affairs, "Interim Measures" [unpublished policy document].

65 *Delgamuukw* v. *British Columbia* [1997] 3 S.C.R. 1010 (hereafter "*Delgamuukw*").

66 Aboriginal law in Canada is undefined because it is determined on a case-by-case basis. Only recently, after abolishment of the Indian Act provision prohibiting land claims litigation, have general principles been developed.

67 Constitution Act, 1982, subsection 35(1).

68 According to the Supreme Court of Canada, Section 35 arises from the fact that First Nations lived on the land in distinct societies prior to the arrival of Europeans. See *R.* v. *Van der Peet* [1996] 2 S.C.R. 407 (hereafter "*Van der Peet*"), 537-48; *R.* v. *Gladstone* [1996] 2 S.C.R. 723 (hereafter "*Gladstone*"); *R.* v. *Adams* [1996] 3 S.C.R. 101 (hereafter "*Adams*"); and *Delgamuukw*.

69 For some First Nations in Canada, the responsibility is extended to the seventh generation.

70 *Delgamuukw*, paras. 125-28.

71 The Supreme Court held that this special relationship may not be severed in such a way as to prevent future generations from holding the same relationship. *Delgamuukw*, para. 126.

72 Ibid.

73 Ibid., paras. 84-108. Until this decision, and the decision in *Van der Peet*, First Nations evidence was not given appropriate weight in court (see the Supreme Court of Canada's discussion of the BC Supreme Court's treatment of oral evidence in *Delgamuukw*). The Supreme Court's ruling in *Delgamuukw* should assist First Nations in bringing women's and spiritual uses of the forests before the courts, both of which do not currently have adequate protection in law.

74 The Supreme Court of Canada recognized that there are instances in which First Nations occupation and use were disrupted "as a result of the unwillingness of European colonizers to recognize Aboriginal title." *Delgamuukw*, paras. 152-53. Presumably, the court was referring to instances when First Nations were dispossessed of traditional territories through obvious means such as forced relocation and flooding of village sites and traditional territories or perhaps through less obvious means such as resource extraction.

75 These are instances in which First Nations shared lands with other First Nations. Until *Delgamuukw*, it was necessary for First Nations to show that they occupied and used lands to the exclusion of other First Nations. *Delgamuukw*, paras. 152-59.

76 Ibid., paras. 166, 169.

77 The Supreme Court of Canada stated that the government must demonstrate that it has accommodated the participation of First Nations in the development of resources in British Columbia through reduced licensing fees, for example. *Delgamuukw*, para. 167.

78 The concept of fiduciary duty did not originate with the *Delgamuukw* decision, but the Supreme Court offered some important clarifications of Aboriginal title in regard to lands under Aboriginal title. The concept of the Crown owing First Nations a fiduciary duty was first enunciated in *Guerin* v. *The Queen* [1984] 2 S.C.R. 335.

79 *Delgamuukw*, para. 168.

80 *MacMillan Bloedel* v. *Mullin [B.C.]* [1985] 3 W.W.R. 577 at 593 (B.C.C.A.) (hereafter *"MacMillan Bloedel"*).

81 [1888] 14 AC 46 JCPC (hereafter *"St. Catherine's Milling"*).

82 *Calder* v. *Attorney-General of British Columbia* [1973] S.C.R. 313.

83 *Haida Nation et al.* v. *British Columbia (Ministry of Forests)* [1998] 1 C.N.L.R. 98 (B.C.C.A.) (hereafter *"Haida Nation et al."*).

84 Recall that Aboriginal title is an exclusive right to the land itself and the resources of that land. It is impossible to reconcile two competing, exclusive rights to the trees on lands under Aboriginal title.

85 As the chief justice of the Supreme Court of Canada noted, "this litigation has been both long and expensive, not only in economic but in human terms as well. By ordering a new trial, I do not necessarily encourage the parties to proceed to litigation and to settle their dispute through the courts." *Delgamuukw*, para. 186.

86 Ibid.

87 It is essential to know that the Supreme Court held in *Delgamuukw* that Aboriginal title is an Aboriginal right but that Aboriginal rights cover a spectrum from general rights to site-specific rights and that this spectrum ends with Aboriginal title.

88 See *R.* v. *Peter Paul* [1998] 3 C.N.L.R. 221 (N.B.C.A.); [1997] 4 C.N.L.R. 221 (N.B. Prov. Ct.); *Thomas Paul* v. *R.*, Victoria Registry No. 981858 (B.C.S.C.); *R.* v. *Chief Dan Wilson et al.* [1999] Vernon Registry No. 23911, unreported decision, J. Sigurdson, 12 November 1999 (S.C.C.); *R.* v. *Chief Ronnie Jules et al.* [1999] Vernon Registry No. 23911, unreported decision, J. Sigurdson, 12 November 1999 (S.C.C.); and *R.* v. *Chief Ron Derrickson et al.* [1999] Kelowna Registry No. 46440.

89 As noted earlier, Aboriginal rights were entrenched in Section 35 of the Constitution Act, 1982.

90 *Van der Peet.*

91 *Gladstone.*

92 *Marshall* v. *R.* [1999] File No. 26014, unreported Supreme Court of Canada decision, 17 September 1999 and 17 November 1999 (hereafter *"Marshall"*).

93 See G. Dawson, *On the Haida Indians of the Queen Charlotte Islands* (Montreal: n.p., 1880), 145-46; and A.P. Niblack, *The Coast Indians of Southern Alaska and Northern British Columbia* (Washington, DC: Smithsonian Institution, 1888), 294-96.

94 Essentially, First Nations will have to show that they used the forests beyond their own needs (no "internal limit") for the purpose of trading items such as canoes with other First Nations. They will also have to show that this trading was an integral part of their culture and "truly made the culture what it is." *Van der Peet*, para. 55.

95 *MacMillan Bloedel*, 584.

96 *Haida Nation et al.*

97 The Klahoose First Nation used a lawsuit to prevent the approval of logging plans for Forbes Bay, an area that the nation had identified as a priority in its treaty negotiations. The Ministry of Forests capitulated to the Klahoose demands before the lawsuit went to court.

98 *MacMillan Bloedel*, 584.

99 Courts have repeatedly held that the federal government owes a fiduciary duty to First Nations to handle their affairs in an honourable manner. This duty applies to the provincial governments as well.

100 In *Adams*, the court considered that in Quebec First Nations could exercise an Aboriginal right to fish for food at the discretion of the minister. The Supreme Court of Canada held that the regulatory scheme infringed on Aboriginal rights to fish. In *Gladstone*, the Supreme Court held that the entire management regime for herring spawn on kelp must be scrutinized because the regulatory scheme also infringed on Aboriginal rights. In both cases, allocation of the fishery did not take into account the existence and importance of Aboriginal rights. According to the Supreme Court, the government must take into account a number of considerations when assessing the existence and importance of Aboriginal rights: whether the government has accommodated the exercise of the Aboriginal right to participate in the fishery, the priority of Aboriginal rights holders, the extent of participation in the fishery of Aboriginal rights holders relative to their percentage of the population, whether the government has accommodated different Aboriginal rights in a particular fishery, how important the fishery is to the economic and material well-being of the First Nation in question, and the criteria taken into account by the government in, for example, allocating commercial licences among different users.
As discussed above, First Nations do not have legal access to forests for traditional uses except with special-use permits. There is no legislative provision to permit access to trees for commercial purposes, such as totem poles carved for sale. First Nations must purchase trees for these uses and for other personal and traditional (noncommercial) uses.

101 *Marshall*, para. 33.

102 *Adams*, para. 54.

103 R.S.B.C. 1996, c. 157.

104 As discussed earlier, First Nations may only obtain special-use permits, issued at the discretion of district managers of the Ministry of Forests.

105 R.S.B.C. 1996, c. 159.

106 See the discussion in the "Aboriginal Title" section regarding the inherent limit.

107 *Marshall*, paras. 29, 40.

108 In another Supreme Court of Canada decision, the court held that the conservation mandate is not only to conserve fish but also to increase the fish stocks. *See R. v. Nikal* [1996] 133 D.L.R. (4th) 658 at 692: The "need to manage the stock goes far further than simply preventing the elimination of the salmon. Management imports a duty to maintain and increase reasonably the resource."

109 It is noteworthy that the 1998 decision to ban coho fishing was made after the Neskonlith Indian Band applied to the Federal Court to implement catch-and-release regulations in the sport-fishing industry to protect the endangered Thompson River coho. *Chief Arthur Manuel et al. v. The Attorney General of Canada et al.* [22 September 1997] FCJ No. T-1497-97 (FCTD) [unpublished]. The 1999 salmon conservation measures were announced after the Neskonlith and Adams Lake Indian Bands filed another action challenging a fisheries management plan on the basis that it did not outline effective in-season management, was likely to cause the extinction of Thompson River coho, and was therefore beyond the power of the minister. *Chief Arthur Manuel et al. v. The Attorney General of Canada et al.* [1999] FCJ File No. T-1364-99. Hearing of the application has been adjourned while the parties attempt to reach a negotiated agreement for the protection of Thompson River coho.

110 For an illustrative discussion of the ways in which Aboriginal law has incorporated traditional principles of stewardship and traditional worldviews despite the strong influence of European-based laws, and the opportunity of Aboriginal law to contribute to the development of Canadian law, see J. Borrows, "With or without You: First Nations Law (in Canada)," *McGill Law Journal* 41 (1996): 629.

6

The Multi-Ethnic, Nontimber Forest Workforce in the Pacific Northwest: Reconceiving the Players in Forest Management

Beverly A. Brown

In a remote Pacific Northwest mountain campground, early season recreational campers settle down for the night. Just before nightfall, four or five vans loaded with Latino men pull in; the men pile out and within minutes set up an extensive camp for thirty to fifty people. By dawn, the campground is empty, and the men are out on the slopes planting trees. Farther north in little coastal towns, poor and working-class white people with pick-ups full of bundled fern fronds stand around in buying sheds waiting to sell their day's work. In the fall, Southeast Asian harvesters of wild mushrooms in the forests of central Oregon set up compact temporary communities, reminiscent of refugee camps along the Mekong River. Throughout the year, altercations break out between non-Native harvesters and Native Americans over routine trespasses on traditional tribal gathering areas in the forest.

Out of the woods, a community-based forestry movement builds national momentum. The movement works in partnerships, collaborative groups, and/or adaptive management area-inspired forums, pioneered primarily by middle-class European Americans who may be long-term residents or, just as likely, recent urban refugees. These groups bring together environmental, industry, rural development, and small business interests to advocate a local decision-making role in forest management on federal lands. Collaborative groups usually solicit participation from all stakeholders, including poor and working-class residents. According to people within those networks, the participation of non-middle-class individuals is rare.

The culturally diverse, highly mobile tree planters, brush pickers, mushroom harvesters, and traditional gatherers referred to above represent a public that defies simple categories of local interests. Over the past few decades, a complex mix of ethnic groups has developed an invisible presence in forest work. Because of ethnic prejudices in rural areas, a desire to live close to culturally intact immigrant families, or a wish for children to attend better schools, forest workers/harvesters from all cultural backgrounds maintain permanent residences in both urban and rural areas. As such, in many community-based forestry scenarios, they are considered nonlocal actors and therefore not stakeholders.

Wherever extensive forests thrive in the world, nontimber forest workers will be found legally or illegally incorporating income from forest activities into their livelihood strategies (Arora 1999; Brown 1995; Fortmann 1995; Fortmann and Bruce 1988; Guha 1990; Hecht and Cockburn 1989; Lynch and Talbott 1995; Peluso 1992; Poffenberger and McGean 1996). In an era when cheap contingent labour has become the cost-cutting goal of industry throughout North America (Callaghan and Hartmann 1992; Hiatt and Rhinehart 1994), the structure of forest work has been rearranged into increasingly temporary, low-wage jobs that undermine the well-being of workers – local or mobile – as well as forest communities. Other than the minimal requirements of labour law, which are only sometimes honoured, companies have exclusive rights over workers' wages and working conditions. Independent nontimber forest products (NTFP) harvesters who sell to company buyers have no legal labour protection.

These individuals usually possess little if any land and are frequently played against one another. Nontimber forest workers generally hold scant if any political power; their voices are silent in policy venues. Nonetheless, their everyday manual labour impacts the environment firsthand, and they represent a constituency that embodies the long-term interests of our culturally diverse, postindustrial society. These workers are the core of the forest-floor labour force. Plans for sustainable forests and healthy communities cannot be achieved without including worker participation in ecosystems management decision processes (Jefferson Center 1997). Moreover, because these individuals work on the vast public lands of western North America, this is an issue with environmental and social implications beyond the public forest itself. How we decide to handle the multicultural constituency of our rural workforce mirrors our intentions as a society.

The community-based forestry narrative at the beginning of the twenty-first century tends to lionize small-town efforts to protect the well-being of rural communities against the broad impacts of large industry and national management (Lee 1994; Lead Partnership Group [LPG] meeting notes 1995-99; Northwest Sustainability Working Group [NSWG] meeting notes 1995-97; *Practitioner* 1996). In practice, *local* and *community* tend to be shorthand for the interests of natural resource businesses (Lee 1994; Lee, Field, and Burch 1990), masking deep divides in rural locales. Rural communities in the Pacific Northwest are frequently stereotyped as homogeneously white, Christian, and pro-timber, a view that belies the accomplished cultural transformations in the US Pacific Northwest. Not only have white environmentalists migrated to these communities, but also Latino and Southeast Asian workers have lived and been employed for decades in the same rural towns with people of Native American and European descent.

In this chapter, I analyze the position of the nontimber forest workforce in current, proposed, and speculative structures of forest management. I focus on the US public policy alternative of community-based forestry (the Canadian version of community forestry arises from similar principles but operates within an entirely different tenure system that makes comparisons difficult [CREWS Congress 1999]). I begin with a discussion of nontimber forest workers, the structures of nontimber work, and the stakes that forest workers have in environmental management and community well-being. Following this discussion is an analysis of three alternative systems of labour in the forest and the weaknesses of each. In the final section, I address the potential intersection of labour and natural resource management within a structure of workers' rights. These speculative proposals accommodate worker participation in a revised version of community-based forest management by recognizing a property right to participate. Persistent attention to structures that accommodate the cultural diversity of the forest workforce is essential to hold any workable system together. I will address US concerns primarily but will provide counterpoints from the BC experience.

Who Works in the Woods?

The tasks involved in nontimber work vary widely, but the workforce travels often across a large geographical area (perhaps including Washington, Idaho, Oregon, northern California, and the length and breadth of British Columbia) under uncertain conditions and erratic employment. No one is yet certain of the extent of this workforce, and research on this large mobile population has only begun (Arora 1999; Hansis 1996; Mackie 1994; McLain and Jones 1996; McLain, Shannon, and Christensen 1995; Richards and Creary 1996; Robinson 1994; Soukhaphonh 1997; Tovares 1995; Yimsut 1995a, 1995b). We know that thousands of nontimber forest workers from Southeast Asian, Latino, Native American, and European American communities labour every year in the forests of the western United States. Punjabi, Vietnamese, Native Canadians, and Canadians of European descent are among the majority of nontimber workers in British Columbia. Across the international Pacific Northwest, nontimber forest workers speak ten major languages – English, Spanish, Cambodian, Laotian, Hmong, Mien, Russian, Vietnamese, Punjabi, and French – and a dozen or more less common languages.

Many immigrants from Mexico, Central America, Southeast Asia, and India lived in the rural backwaters of their home countries, but today they are more likely to live in cities or large towns in the United States or Canada. Some have been displaced by extreme poverty and regional conflict (as in the case of Latino and East Indian immigrants), while others have been forced from their homes by the aftermath of the Vietnam War or Cambodia's genocidal Khmer Rouge (as in the case of Southeast Asian refugees). Many immigrants in forest work are familiar with agricultural activities, speak English poorly if at all, and do not read or write in their original languages. Southeast Asian and Guatemalan refugees suffer bouts of post-traumatic stress after having witnessed the murders of their friends and relatives in their countries of origin. Upon arriving in the United States, people's first priority is simply to survive in order to provide for their families. Consequently, people agree to work under conditions that would not be acceptable to most people born and raised in the United States or Canada.

European-descent and Native nontimber forest workers reside in small cities as well as rural towns and hamlets. Two tiers of workers divide unskilled and skilled tasks, one at low income, the other in the lower-middle income range. Many Anglo and Native harvesters supplement marginal livelihoods, while skilled technical survey microcontractors and firefighters can pull down good money when they obtain work. Most Anglo and Native workers are literate in English.

Structure of Work

Two sets of people seek nontimber forest work. First, there are *people who want to work in the woods*, whether they are unskilled or skilled in forest-related tasks. This group includes individuals from all ethnic backgrounds who have a long-term commitment in forest employment. Among this group, people of European descent have the greatest access to viable, reasonably well-paid forest occupations. Incomes range from poverty level to lower middle class. In British Columbia, this group includes thousands of university students who plant trees during spring and summer. Second, there are *people who must find any kind of work*. This group includes poor people in general, many first- and second-generation immigrants, and rural in-place Native and European-descent individuals who may or may not have an interest in forest work over time. Clearly, these sets of workers are not exclusive: people who start out working in the woods out of economic desperation may come to love the work and choose forest activities as their occupations.

In the United States, four systems of work relations characterize nontimber employment. First, some crews work out a labour-based service contract. The contractor must post high bonds and have up-front capital to float the organization and provision the job and obtain the contract through low-bid competition. Tree planting is usually organized in this way. Second, and less often, individuals microcontract for semiskilled or skilled technical surveys. Stream-side restoration and vegetative surveys will sometimes fall into this category. Third, individuals or informal crews will harvest under a person or company that has obtained an exclusive licence to a specific large parcel of land from private industry or a public agency. This is typical of floral greens harvesting in southwest Washington and is a growing practice elsewhere. And fourth, individuals and kinship groups may buy permits, harvest as independent businesspeople, and sell as individuals to company-hired buyers. Independent work typifies mushroom, medicinal herb, and some floral greens harvesting.

Nontimber forest workers engage in two kinds of activities. Forest contract labour involves tree planting, brush cutting, and thinning. Forest-floor workers harvest nontimber forest products, including mushrooms, floral greens, and medicinal herbs. Some activities have become partly segregated; for instance, tree planting and brushing are primarily low-income Latino activities (in contrast, in Canada these workers are primarily well-educated European Canadian members of the "counterculture"). A high proportion of matsutake mushroom and beargrass harvesters are of Southeast Asian descent. Crews or kin/friendship groups almost always segregate by ethnicity and language.

Forest contract labour and harvesting of NTFPs are usually considered to be two completely different types of work, but they are closely intertwined in several ways. First, both contract labour and commercial harvesting depend on mobile, contingent workers who are not permanent employees of any agency or industry. Second, the increasing use of licensees as intermediaries in nontimber harvesting mimics the contract system. Contractors and licensees have considerable power over crews because there is virtually no oversight from outside parties and many incentives to cut costs on the backs of the workers. Landowners, public and private, deny responsibility for personnel relations between contractors/licensees and workers. Third, based on forest-floor activities, both systems of work impact the environment on a daily basis. Harvesting methods can affect a multitude of biological niches, and the methods by which contract service work takes place will affect harvesting outcomes and long-term ecosystem functions. Fourth, as noted above, workers participate in multiple contract and/or harvesting activities during the year, so that, from a worker's point of view, the systems are related in patterns of livelihood strategies. And fifth, US community-based forestry and stewardship contract scenarios proposed by collaborative groups and/or industry combine various aspects of both systems (DeConcini 1994; LaRocco/Fazio/English 1994; United States Forest Service 1992; Williams 1995; also LPG and NSWG meeting notes and personal communication with Applegate Partnership and others 1995-99).

As long as timber provided good incomes for local businesses and the blue-collar middle class, the low-wage nontimber tasks such as tree planting, brushing, and nontimber products harvesting were considered marginalized, low-status work fit for peripheral members of society. Although migrant work and informal economies based on the forest kept many households afloat for years, the level of visibility was low (Brown 1995; Fortmann 1990; Ratner 1984). For instance, although numerous Latino "service" contract workers have travelled through and camped near forest communities for decades, as one community activist said, "We chose not to see them."

Agencies, NTFP buyers, and contractors have been served well by the relative invisibility of nontimber crews working deep in forests. The invisibility of the labourers, coupled with the limited accountability of the contractors, licensees, and buyers, has led to substantial abuse of workers (Jefferson Center 1996; Mackie 1994; Tovares 1995; Vaupel and Martin 1986). The way to cut costs in competitive-bid and/or labour-based forest work is simply to pay people less money, and workers report that the current system depresses wages and working conditions in forest-floor occupations, even when overt abuse is not present.

During any one year, forest workers might work at one or all of these nontimber tasks. They might supplement their incomes through seasonal work in agriculture, home microbusiness, logging, seafood processing, landscaping, or tourism or through welfare and/or the informal economy. As such, tens of thousands of forest workers are already part of the contingent workforce that increasingly characterizes much of North American employment. Few contingent systems, and none of the US-based nontimber forest work systems, provide any form of health insurance, sick days, vacations, or other benefits. Canadian tree planters are fortunate to have access to a national health care system. In the United States, legally mandated benefits, such as workers' compensation and social security, are sometimes sidestepped by contractors who list only portions of their workforces on legal documents. NTFP harvesters are considered independent and thus have no protections. Canadians report far fewer abuses of the workforce, due in part to organized efforts to improve working conditions and camp standards.

Company Contractors/Employees, Community Participants, or Occupationally Organized Workers?

In the United States, the current mix of woods work systems serves the interests of forest-related companies very well and mirrors a traditional employee-employer relationship. Companies hold exclusive decision-making power, including the essentially unchecked right to hire and fire workers at will. Challenges to unfair labour practices usually rest on the shoulders of individual workers, and legal challenges involve considerable initiative, time, and money. Because of their vulnerable position in the system – since independent NTFP workers have no employment rights at all – workers who depend on access to contingent forest work are unlikely to become involved in controversial public dialogues about forest management, unless they defend an industry line.

Although workers have not yet found a way to challenge the power of business in the forest industry, rural community development activists and local governments have balked at this power over the past decade. Discussed at some length below, the community-based forestry movement arose in response to the loss of jobs, revenue, and infrastructure that followed the severe decline of public forest logging due to a variety of industry-initiated business decisions and environmental factors on US public lands at the end of the 1980s.

In most US community forest scenarios (of which there is a wide variety from "right" to "left"), workers are talked about at great length as community members who need jobs to remain functioning parts of the community. As such, community-based forestry advocates strongly for workers, but within two models. The first situates workers as employees with a wider range of skills to engage in year-round forestry labour and thus with more stable jobs. The second places workers as independent businesses, usually as contractors (who would then hire within the traditional system of employment). The heart of most community-based forest models is the empowerment of local businesses, mediated by local government or analogous new institutions, to participate in the decision-making process on when and how natural resources are managed in a locality.

Workers' opportunities to shape forest policy and management are limited. In the United States, they cannot easily improve their position in the traditional industry-favoured labour system, nor, for a variety of reasons, are they easily able to join community-development and local business advocates. Alternatively, occupational associations have the potential to influence dialogue, but at the risk of workers' jobs. Although such associations have no real power, they are essential in bringing workers' voices to the forest dialogue. Real change requires workers to take the next step and form a legally recognized negotiating entity. The difference between the United States and Canada in this regard may bring about radically different outcomes in forest management.

Forest workers reside uneasily in a system in which labour has few rights and even less public status. The right to form a business and organize capital is well respected. The ability of labour to organize in response to business has diminished drastically in the United States (as elsewhere on the planet). Under the current National Labor Relations Board (NLRB) interpretation and judicial enforcement of US labour law, virtually none of the contingent forest workers would realistically be able to organize a union (Green 1990; Gross 1996; Pavy 1994; Rosenblum 1995). In Canada, where organized labour still retains effective rights (as opposed to on-paper rights) of organizing, the potential for forest workers to gain some leverage in negotiating the terms of their work is substantially better (CREWS Congress 1999). However, even in Canada, the burgeoning practice of hiring temporary "just-in-time" workforces presents a challenge.

Labour, Forests, and Community Together, Not Separate

Communities are part of forest watershed ecosystems, and all forest management decisions have some form of direct impact on people as well as trees. Workers are the backbone of forest-related communities, upstream or far downstream in the city. Forest workers in particular have intimate knowledge of everyday forest management.

Workers impact and are impacted by forest management decisions in the workplace, in leisure activities, in their pocketbooks, and at their homes. Industry-favoured systems of labour, the current community forest movement, and traditional labour organizing all tend to separate workplace issues from larger resource management issues. The dismissive refrain "Oh, it's just a labour issue" can be heard time and again in forest dialogues when the issue of forest workers comes to the table.

In this chapter, I do not discuss the positions of environmental groups regarding forest workers. This is largely because environmental groups, with rare exceptions, have not bothered to think seriously about the inevitability of forest labour under any scenario, highly managed or zero cut. Given the public nature of forests and waterways, one of the great challenges of US forest management in the new century is to bring labour and natural resource issues within the shared vision of democratic participation. Specifically, how can ethnically and culturally diverse community members who are also forest employees, or otherwise attached to a forest-related place of business, gain effective rights to participate in forest management? More importantly, how can they gain those rights without fearing retaliation by an employer and/or harassment based on cultural stereotypes or xenophobia?

Traditional industry-favoured labour systems have essentially no interest in workers' participation that contradicts the industry agenda. I will not pursue that subject further here; however, US community-based forestry has developed a set of ideas that, although focused on business and treating labour as essentially passive, may open the door to different ways of thinking.

From Marginalized to Central in Community-Based Forestry

When mills in the Pacific Northwest began to close in response to mechanization, industrial restructuring, and to some extent the decline of logs from public lands due to environmental restraints, rural community development advocates looked for ways to replace lost timber jobs with something equivalent. Wood products workers and businesspeople alike expressed bitterness about the ability of timber companies to pick up and leave, stranding those who had become dependent on them. Some people lashed out at the urban-based environmental lobbies. Others, attempting to diminish a locale's vulnerability to the boardroom decisions of international industry, sought opportunities in the local landscape and economy.

Community development advocates and industry quickly realized that ecosystems management would include the total set of activities taking place in a forest. *Nontimber forest activities no longer seemed to be marginal at all – they came to be seen as central to a comprehensive understanding of ecosystems management.*

Ecosystems management rose to prominence about the same time that the NTFP industry began to diversify, streamline, and take off. Several researchers speculate that the value of NTFP over eighty years may exceed the value of an eighty-year timber rotation (Yvonne Everett, Hayfork, California, personal communication, 1996). If timber thinning were added to the NTFP value, then an economic strategy integrating the forest floor with the forest canopy might provide the best economic returns over time.

The backbone of US community-based forestry efforts is collaborative groups, partnerships, or groups that are functional equivalents. Collaborative groups strive to come to mutual agreement on a set of topics and are willing to put aside subjects on which they cannot agree. They do not currently possess decision-making power, but they do initiate projects for which they seek federal and private landowner approval and/or participation. Partnerships are local by design. Some, such as the highly lauded Applegate Partnership, overlap with the adaptive management areas of President Clinton's Option 9 Forest Plan. Others, such as the Lead Partnership Group, bring together bioregional groups and practitioner groups focusing on locally based, project-specific work. They interact with national efforts through several US-based networks, including the National Network of Forest Practitioners, the Communities Committee of the Seventh American Forest Congress, and the Pinchot Institute. American Forests stays in contact with the various groups, networking people among all forest-related interests at the Washington, DC, level. Collaborative groups offer the most widely accessible forum in which stakeholders in forest management can share information and creative thinking, even when collaborative partners disagree with or distrust one another.

The twin interests of science and society – site-specific ecosystems and locale-specific community-based forestry – have focused the deliberations of collaborative groups. Mechanisms suggested by such groups vary, but often they include proposals for some form of multiyear stewardship contract serving both science and society, including (1) consistency of applications and monitoring focused on specific ecosystems over a multiyear period and (2) more predictable supply of income for businesses and employment for workers, with the result of well-being for the community.

National forest products companies occasionally participate in collaborative groups and have developed an interest in "stewardship" arrangements consistent with longer-term, multispecies management as proposed by community forestry (LPG conversations with industry representatives 1995-99). But this interest may result from political calculations rather than from a commitment to rural communities. In one of the collaborative meetings, a representative of a large timber company figured that branch operations putting out their shingles as local companies would probably be included in a local preference arrangement.

Community development groups wield significantly less influence than industry, but they put forward powerful social arguments for community well-being. Inclusion of community well-being objectives adds social legitimacy with which profit-driven industry struggles to compete. Some community-based forestry partnerships promote the statutory creation of collaborative "stewardship councils" that co-coordinate forest management activities with the federal forest agencies (with the level of decision making undetermined). These councils might diffuse industry power and steer multiyear contracts in the direction of locally based business. A 1995 bill introduced by Representative Williams of Montana sketched the kind of system that would in effect put the power of the law behind local preference (Williams 1995).

National environmental groups oppose any form of collaborative forest management, arguing that, whatever the intent, organized financial power will likely dominate collaborative forums and dictate the final outcomes. Since most collaborative groups pivot their cases on more (but better) business and regional development revenues, the national environmental groups fear that rural partnerships with any level of effective power would eventually promote greater resource extraction or activities that lead to increased environmental impacts (McCloskey 1995).

National environmental groups derive the bulk of their influence through large urban constituencies and contributions. As watchdogs over agencies and industry through administrative and judicial review, the environmental groups focus on broad principles that may or may not be helpful in site-specific ecosystem situations. How environmental groups propose to be involved in local forest dilemmas that do not fit general models is not always clear, but a key element appears to be in instituting "all-party" monitoring systems that could establish a multitude of eyes on the conditions of local forests.

What power would workers have in any of these scenarios? Significantly, there would be very little leverage for workers in the local preference proposals asserted by community forestry advocates. In the case of federal lands, compliance with federal minimum wages for forest tasks might be better enforced at the local level. But nothing much would change, because the traditional labour system would continue to go unchecked. Whether local or nonlocal, workers' participation in setting the terms and conditions of work, including participation in ecosystems management, would rest on the unilateral good will of their employers and/or NTFP companies, large and small. Workers who do not conform to narrow "local" requirements and who, because of ethnic/racial and/or linguistic differences, may not be welcome in a particular "local" area, could exert no leverage at all in a local preference system. Again, because of the dense patronage systems in resource work and rural communities, local workers would continue to hesitate to participate in forest management issues off the job, largely from fear of company retaliation. This is a pivotal issue that seems to be opaque to most well-educated middle-class reformers of any persuasion and to blue-collar workers.

Community-Based Forestry and Stewardship Contract Proposals

Between 1990 and 1997, proposals for various forms of "stewardship contract" (under this and other names) were proposed in Congress (Ringgold and Mitsos 1996). Stewardship contracts were conceived as multitask and multiyear arrangements. Some of them instituted goods-for-services contracts, providing an exchange of restoration work for logs (United States Forest Service 1992). Others are types of local preference legislation that would give favour to local forest-related businesses in bidding for labour-based "service" contracts on public land (see Williams 1995) and access to NTFP. Multiyear contracts, whether community-based or industry-based forestry, may be extremely beneficial in ecosystems management, bringing needed continuity to environmental tasks and providing some economic stability to businesses and their employees. However, they are essentially a *turf-based* method of seeking protection of sectoral interests. In 1998, the US Congress authorized twenty-eight stewardship pilot projects, mostly in the western states (United States Congress 1998).

Industry-preferred proposals include multiyear contracts on large parcels. Such contracts are cheaper administratively (in the short term) for both federal forest agencies and private industrial landowners. Contractors might be local or nonlocal. Because the contracts are competitive, winning bidders would probably continue to choose the cheapest workers who can accomplish adequate work (except in some high-publicity demonstration projects), thus maintaining a preference for low-wage labour. For business efficiency and cost reduction, work tasks would likely remain specialized and partly segregated. Land managers (both public and private) and their contractor intermediaries would retain all prerogatives in relation to workers. The ethnic diversity of workers would remain high, wages and working conditions would likely remain low, and ethnic tensions would remain exploitable.

Conversely, the community-based forestry movement prefers stewardship contracts that include smaller parcels and/or local preferences. This scenario gives local businesses an advantage in bidding for contracts and/or licences to harvesting sites (some proposals allow only small businesses – twenty-five employees or fewer). For the locality, there are two major advantages. (1) Local business would generate cash flow in the area, including economic activity that would benefit the funding of retailers, schools, and the local government. (2) Local workers would, theoretically, have a preference for work. The optimal community forestry vision is the creation of independent local businesses that acquire contracts/licences – maybe leases – that would provide family-wage jobs based on the multi-task work of semiskilled and skilled local forest technicians. Workers would perform a variety of activities throughout the year and replace the task-specialized mobile workforces of the past (Jobs-in-the-Woods documents; LPG conversations; personal communication with community development advocates 1994-99). Local business owners would gain considerably; however, as is the case with industry preference schemes, employee-employer relationships would not be altered much if at all from traditional labour systems. Workers would still be subject to complete managerial discretion, and there is no guarantee that they would be treated better than in the former system. (In rare cases, such as cooperatives and socially conscious enterprises, working conditions and wages would likely be much improved.) Overall, the ethnic diversity of workers would likely go down (due to local political advantage and xenophobia). The effect on wages and working conditions would be unstable, because any locality is embedded in a vastly larger economic system that responds to goods and labour pricing across an entire industry, not in a local area.

Which Kinds of Property Claims?

The property implications of community forestry are currently limited to a narrow vision based on geographic boundaries. Statutory preference for local businesses – mediated by stewardship councils or a local government – could be conceived as conferring a kind of property right to a bounded geographic area. Such an arrangement would extend the power of the business elites, who already dominate many rural towns, to adjacent federal forests. Who would be included? Who would be excluded? What are the checks and balances?

Multiyear contracts with wide responsibilities on specific parcels of land invite investments of time, equipment, and training by the businesses granted contracts/licences. With renewed contracts, these businesses – or local preference stewardship councils or their equivalent – could conceivably begin to establish limited property rights over the land itself. Such "property" would be a "bundle of rights," legally recognized over a piece of land but not necessarily connected to the ownership of the land. Such a bundle that constitutes legal property over nontangibles processes or possessions can create significant discretion and influence; for instance, Reich (1964, 1990) has argued that medical licences constitute this kind of abstract property. One could argue that certain kinds of rights to long-term decision making in multiyear contracts could be interpreted as property rights in themselves.

There is a precedent for the evolution of contracts into quasi-property and full property rights. Possessors of long-term stewardship contracts could, as has been done with grazing leases (Calef 1979; Klyza 1996; Nelson 1995), argue for multiyear property-type rights over otherwise public lands. The inherent property implications of local preference regimes that base community protection on turf-based proposals raise legitimate questions about which property claims might evolve from them and in whose interest. To ignore the potential property implications is naive.

Which rights might workers establish in a turf-based system of multiyear contracts? Unless the businesses are worker owned, the protection inherent in turf-based contract systems devolves to a business entity, not to employees, as either workers or local citizens. Conditions and terms of work, including workers' participation in the everyday decisions of ecosystems management, would be solely at the discretion of the employer. Would small businesses be any more considerate of employees' rights? Not necessarily, and not in the experiences of many employees in small businesses, who are subject to more arbitrary management than employees of large companies, in which various procedures for recourse are in place to smooth large hierarchical operations (especially when a union protects the interests of a diverse workforce). The participation of an ethnically diverse workforce would be subject to local prejudices; sadly, our long national histories in the United States and Canada demonstrate that companies are not above recruiting ethnically different workforces that can be played against one another in a locality.

Occupational Organizing

In contrast to the experiments in community forestry or industry preference models of traditional employer-employee relations, recent US training programs for forest workers, occupational associations, and union models have remained relatively static over the past ten years. Jobs-in-the-

Woods (later called Ecosystems Workforce Training) programs in the United States were part of the Economic Adjustment Initiative attached to President Clinton's Northwest Forest Plan. Designed with former timber workers in mind, the actual programs provided training to a variety of displaced workers. In Oregon and northern California, these programs were conceptually related to community forestry with the intention of creating more local year-round, family-wage jobs in the forests. People were trained for a set of high-end labour skills for an ecosystem restoration industry that so far has not materialized. In one vision of the training program design, high-end, year-round workers replace "migrant" (which in practice meant many culturally diverse workers), single-task, short-term crews. In a subsequent vision, the highly skilled workforce was championed to the neglect of the large numbers of lower-wage, labour-intensive workers who – often immigrants or other people working their way out of poverty – work extremely hard, do a narrow set of tasks very well, are cheaper, and therefore preferred by contractors. Recent talk about unionizing the high-end restoration workers followed the same pattern. The union was to be only for the restoration, the "craft" part of forest labour (almost exclusively white, mostly male), while the lower-end majority of "industrial" workers (from diverse ethnic backgrounds) would be left to fend for themselves.

Union preferences for high-end workers over low-end workers and training program preferences for "local" workers over "mobile" workers correspond with the weaknesses of the community-based forestry initiatives. Preferences for local or skilled workers create a kind of exclusionary social and geographic zoning of forest management and labour. Past experiences with exclusionary zoning in urban areas (Davis 1991; Plotkin 1987), and the track record of racial/ethnic inequality in most rural areas (Hanks 1987; Zekeri 1995), suggest that given the opportunity local elites (usually, in the Pacific Northwest, well-off members of the white, Christian majority) participating in community forestry programs may end up, by obliviousness or design, supporting discrimination.

Of course, the mandate for inclusion of the rights of a mobile and ethnically diverse workforce originates in the fact that we are already multicultural, highly mobile, postindustrial societies in Canada and the United States. The future of the latter country as a democratic nation depends on how well its citizens negotiate their way toward a fully inclusive society in cities, the suburbs, and the countryside. To re-create in rural areas the turf-based system of income and racial exclusion that so plagues our urban areas is morally repugnant.

An Alternative Proposal
Nontimber forest workers in two initiatives, one US, one Canadian, recently began to address some of the structural challenges to include

diverse workers across the nontimber forest industries. In the United States, the Alliance of Forest Workers and Harvesters became an independent organization in 1999. The alliance is a multicultural association of both skilled and nonskilled forest workers, mobile and "in-place," across "service" contract and NTFP harvesting. The organization was structurally designed from the beginning to represent the wide diversity of cultural and ethnic groups working in the woods (Jefferson Center 1998, 1999). In Canada, bottom-of-the-line labourers – the tree planters – created the Canadian Reforestation and Environmental Workers Society (CREWS) and, while being courted by two union groups, are deciding whether to pursue unionization. As the name implies, CREWS embraces reforestation- and ecosystems-related workers across the industry. While CREWS is not sure how to deal with the fluidity of the contingent forest workforce, it is facing that issue head on as it stretches to envision a new concept of organizing (CREWS Congress 1999).

These two organizations are taking bold and essential steps forward and have established active communication with each other across the border. However, aside from potential successes in achieving inclusivity across the multicultural mobile and local workforce or attaining legally binding collective rights of negotiation on labour issues, how will they gain direct participation in the forest management decision-making process?

A Property Right to Be Involved in a Process

The community-based forestry movement has argued for communities in or adjacent to forests to have some right of participation in discussions about how forests are to be managed. In spite of the objections that I have raised above, the community-based forestry movement can be credited with reinvigorating the concept that bundles of rights over natural resources are not the exclusive possession of land management entities (public or private) or industry. Such a concept – a technical definition of property – is abstract and refers to a general understanding of the importance of, say, the involvement of localities in forest management. It is not a description of sovereignty over a physical piece of terrain. This abstract idea opens the door to the idea that labour may be one of the subsets of groups that could establish bundles of rights in relation to the process of forest management.

Both commercial and public land managers command a decision-making position by default. Industry controls a large chunk of money, and land managers (public and private) control a recognized bundle of rights over a set of natural resources. Both bundles of rights, although different, can be understood as property rights. Community forestry advocates are knocking at the door with a proposed bundle of rights of their own, based on proximity and economic connections, to become one of the groups with decision-making leverage in forest management. The bundle of rights that they would like to acquire is a kind of property right, but it need not be associated, as is generally now advocated, directly with a piece of geography with a line around it. The idea could be much more abstract, legally legitimizing the "right" of a locality to be involved in natural resource issues (e.g., a community far downstream may have a legally legitimate interest in the management of a particular watershed). Many environmentalists would decry the concept of such a bundle of rights extending beyond government or private property mediated by government regulations (with some exceptions, e.g., one environmental offshoot encouraging environmentalist-run public land trusts [United States Senate 1995]). Most environmental groups demand that the de facto property rights of industry and public land managers be adjusted only through judicial or legislative means and would oppose any other group attaining a new bundle of "property" rights in the management of forests.

Two quite different models are in contention here. One model recognizes that industry (in possession of the bundle of rights over capital) and private or public land managers (in possession of the bundle of rights over natural resources) possess an inevitable bundle of rights that have effects on forest management. In this model, though, additional groups should not be recognized. To change policies or procedures, actions of industry and public land managers must be challenged through the courts and Congress. The other model suggests that various entities, with different but related bundles of rights, can create a diverse set of legitimate participants with a kind of "property" right in the decision-making process – not the land/forest itself (and any decisions would still be subject to all the challenges in courts and Congress).

In the first case, nontimber labour, if it were organized, could of course become involved in traditional lobbying. In the second scenario, labour itself could, for the duration of a contract (and it would be a new type of contract), become one of a larger set of groups that gained a bundle of rights in the decision-making process. This would be, in other words, labour's *property right to participate in the process of forest management.*

Rather than being a turf-based concept of geographic preference, such as community forestry often tends to be, or a concept of a local (or locally organized) workforce as envisioned by retraining programs, this idea omits the bounded geographical idea of turf. The idea rests solely on an abstract "property" bundle of rights, not in any way a line around a particular parcel. For labour, such a concept could avoid the danger of workers organizing by ethnic locality or only in a craft model that excludes other parts of the industry. For instance, a broadly based coalition of labour entities, representing mobile and local, ethnically diverse, high- and low-skilled labourers, could overcome two challenges. First, labour would be organized across diverse lines, so businesses could not leverage "local" versus "non-local" to stir up ethnic or other resentments that ultimately pit workers against one another and eventually lower wages and working conditions. Second, the separate strengths of mobile and in-place workforces could be realistically evaluated in light of the bigger picture of forest management in an ecosystem comprising small watersheds and/or mighty rivers extending hundreds of kilometres. Historically, unions have often provided internally funded training (Cobble 1994; Wial 1994), facilitated communication, contract-negotiated grievance procedures, and leveraged for honest discussion of the essential contribution of labour to ecosystems management.

This speculative scenario would bring together labour, community, and forest management into an integrated whole and (by not excluding court, legislative, collaborative, and media challenges in the forest sector) would guarantee legal enforcement of national laws and regional/national NGO participation in forest management. By establishing a "property" right to the process of forest management, labour's presence would be *required* to forge long-term solutions. It would free up workers to participate in that process, which is only a theoretical right at the moment.

Other benefits might come to the fore. Environmental groups might find virtue in the possibilities created by a large environmentally oriented group of workers that no longer avoids civic participation because people no longer fear employers' retaliation. Furthermore, monitoring is one of the keystones of environmental-backed forest experiments. The establishment of scientifically credible monitoring of forest health is time-consuming and extremely expensive (LPG conversations 1996-99). If workers were recognized to have a day-to-day right to and responsibility for first-level monitoring as part of their overall responsibilities, then cost-effective monitoring might be attainable. With the protection of a labour contract including a right to be involved in the management process, workers could participate in monitoring in their off-hours as local and national citizens as well, with considerably less fear of retaliation by employers.

The success of such a hypothetical proposal would depend on the inclusion of migrant and local workers in a fair and democratic union, thus short-circuiting the manipulative leverage of ethnic prejudices and cost-cutting practices (this might be a new kind of union that would not look much like the old unions of the factory-dominated age). Partnerships among several regional forest unions, forming a strong, democratic national union, could provide safeguards for equal opportunities for the culturally diverse individuals – including women – in forest work, especially when in contract relations with a federal forest agency that, in the United States, is mandated to pay attention to equity.

A final condition would be necessary, one that addresses the large number of undocumented forest workers, especially in the Latino community. To achieve a genuinely sustainable system, we must also look to the future, when not only money and goods but also people will travel freely across national borders. Indeed, over the past two decades, working people have been travelling with or without papers across borders all over the Americas and have established themselves as the backbone of forest contract labour. At some point, immigration laws need to catch up with reality rather than impose heavy penalties based on concepts dependent on a less mobile and less thoroughly interconnected world – a view of the world that is long outdated. Migration from Mexico and Central America to the United States is inevitable. Without the freedom to organize across borders, and ultimately to protect the full human and civic rights of workers moving across borders (and not under the servitude-like terms of "guestworker" programs), it will be impossible to institute justice for workers or any sustainable systems of labour in natural resource work.

Just a Speculative Exercise?

Implementing a forest management process as outlined above is purely a theoretical exercise at this writing. Both US and Canadian realities, although different, would not support either the labour or the public participation pieces necessary for success (CREWS Congress 1999; Gross 1996).

Even if it were possible, would forest workers be interested in it? The answer is not clear. Many forest contract workers and commercial NTFP harvesters dislike the unfair conditions under which they also frequently work, but they love the freedom of the woods. They would likely balk unless there were some innovative form of collective bargaining. The experiment that CREWS is conducting in Canada among tree planters will be instructive as they figure out whether to persevere as an association or to organize a new kind of union. Such a union would attempt to unscramble who is labour and who is management in contingent, contract, or subcontract work. This kind of scrambling has been taken to an extreme in some other industries, such as janitorial work, in which subcontracting replaces employees (Howley 1990). Other options are also possible. Cooperatives and cooperatives of cooperatives may appeal to some people – perhaps even cooperatives in some collaborative partnership with unions. Forest workers will have to decide over time which kind of strategy will best serve them – unions, associations, cooperatives, some new configuration, or nothing at all. The only thing certain at the beginning of this century is that nontimber forest workers in the United States have almost no rights, and Canadian nontimber forest workers are uncertain what the future will bring. This is not the situation upon which ecosystems sustainability or community well-being can be built.

Conclusion

Can we even envision a forest management system that, rather than seeing workers as passive recipients of the largesse of business (i.e., providers of jobs), can integrate nontimber (and timber) workers as responsible, independent, and indispensable agents in forest management? Can we envision a system of negotiations among various groups recognized to have bundles of rights (i.e., a kind of property right in the *process* of forest management)? Can we envision a new form of interactive checks and balances, in addition to and hopefully preceding one-size-fits-all national court action, that adds necessary fine-tuning to congressional laws and agency regulations?

Tree planters, mushroom pickers, bough and medicinal plant gatherers – migrant or local, commercial or traditional, in poverty or lower middle income, and of whatever language and ethnicity – are not marginal in the least. They are the everyday life of forest-human relationships, working across the boundaries of formal and informal economies everywhere in our forest systems. These are the people – the public – who constitute healthy communities, and they originate in urban as well as rural places in our intensely mobile society. Turf-protection arrangements, including privatization and overt/covert ethnic segregation, have not protected any of us in the cities. Ecosystem protection will ultimately come from the commitment of ethnically diverse people who have a fair, accessible, legally recognized, and participatory stake in the health of our geographical and biological communities. This right may well include new concepts of property, with a property right to participate in the process of forest management. Meanwhile, promoting the actual right (as opposed to a paper right) for workers to organize, and analyzing the implications of collective organizing among ethnically diverse nontimber forest contract and harvest workers, are crucial steps toward the goal of achieving dynamic sustainability of our collective forests, watersheds, and multicultural society.

Organizations

Alliance of Forest Workers and Harvesters, Olympia, Washington.
American Forests, Washington, DC.
Applegate Partnership, Ruch, Oregon.
Canadian Reforestation and Environmental Workers Society (CREWS), Vancouver, British Columbia.
Collaborative Learning Circle, Ashland, Oregon (formerly a project of the Rogue Institute for Ecology and Economy, Ashland, Oregon), Cate Hartzell, coordinator.
Lead Partnership Group (LPG), Taylorsville, California, Jonathan Kusel, coordinator.
National Network of Forest Practitioners (NNFP), Boston, Massachusetts (formerly a project of Forest Trust, Santa Fe, New Mexico), Thomas Brendler, coordinator.
Northwest Sustainability Working Group (NSWG) (disbanded 1997), Institute for Washington's Future, Seattle, Washington, Don Hopps, coordinator.
Pinchot Institute for Conservation, Washington, DC, Al Samples, director.

Seventh American Forest Congress Communities Committee, Watershed Center, Hayfork, California, Lynn Jungwirth, coordinator.

References

Arora, David. 1999. "The Way of the Wild Mushroom." *California Wild* 52 (4): 8-19.

Brown, Beverly A. 1995. *In Timber Country: Working People's Stories of Environmental Conflict and Urban Flight*. Philadelphia: Temple University Press.

Calef, Wesley. 1979. *Private Grazing and the Public Lands*. New York: Arno Press.

Callaghan, Polly, and Heidi Hartmann. 1992. *Contingent Work: A Chart Book on Part-Time and Temporary Employment*. Washington, DC: Economic Policy Institute.

Cobble, Dorothy S. 1994. "Making Postindustrial Unionism Possible." In Sheldon Friedman et al. (eds.), *Restoring the Promise of American Labor Law*. 285-302. Ithaca, NY: ILR Press.

CREWS Congress. 1999. "Report on Proceedings of Canadian Reforestation and Environmental Workers Society Congress." Centrepoint POB 19644, Vancouver, BC, V5T 4E7.

Davis, Mike. 1991. *City of Quartz: Excavating the Future in Los Angeles*. London: Verso.

DeConcini (Senator). 1994. Stewardship End-Results Contracts Demonstration Act (S. 2100). 103rd C. 2. S. United States Senate.

Fortmann, Louise. 1990. "Locality and Custom: Non-Aboriginal Claims to Customary Usufructuary Rights as a Source of Rural Protest." *Journal of Rural Studies* 6(2): 195-208.

—. 1995. "Talking Claims." Paper presented at the Annual Meeting of the Rural Sociological Society, Portland.

Fortmann, Louise, and John Bruce. 1988. *Whose Trees? Proprietary Dimensions of Forestry*. Boulder, CO: Westview.

Green, Hardy. 1990. *On Strike at Hormel: The Struggle for a Democratic Labor Movement*. Philadelphia: Temple University Press.

Gross, James A. 1996. *Broken Promise: The Subversion of the US Labor Relations Policy, 1947-1994*. Philadelphia: Temple University Press.

Guha, Ramachandra. 1990. *The Unquiet Woods*. Berkeley: University of California Press.

Hanks, Lawrence. 1987. *The Struggle for Black Political Empowerment in Three Georgia Counties*. Knoxville: University of Tennessee Press.

Hansis, Richard. 1996. "The Harvesting of Special Forest Products by Latinos and Southeast Asians in the Pacific Northwest: Preliminary Observations." *Society and Natural Resources* 9(6): 611 ff.

Hecht, Susanna B., and Alexander Cockburn. 1989. *The Fate of the Forest: Developers, Destroyers, and Defenders of the Amazon*. New York: Verso.

Hiatt, P.J., and Lynn Rhinehart. 1994. "The Growing Contingent Work Force: A Challenge for the Future." *Labor Lawyer* 10(2): 143 ff.

Howley, John. 1990. "Justice for Janitors: The Challenge of Organizing in Contract Services." *Labor Research Review* 15: 60-70.

Jefferson Center for Education and Research. 1996. *1996 Forest Worker/Harvester Gathering* (West Salem, OR, November). Wolf Creek, OR: Jefferson Center.

—. 1997. *Letter to Chief Dombeck* (report on the spring 1997 Forest Worker/Harvester Gathering [Redding, CA, March]). Wolf Creek, OR: Jefferson Center.

—. 1998. "Report on the Jefferson Center Forest Worker/Harvester Program." Newsletter. 1 (1-4).

—. 1999. "Report on the Jefferson Center Forest Worker/Harvester Program." Newsletter. 2 (5-6).

Klyza, Christopher M. 1996. *Who Controls Public Lands? Mining, Forestry, and Grazing Policies, 1870-1990*. North Carolina: University of North Carolina Press.

LaRocco/Fazio/English (Representatives). 1994. Stewardship End-Result Contracts Demonstration Act (H.R. 3944). United States House of Representatives.

Lead Partnership Group. 1997. *Blairsden Report* (report on 1994 meeting in Blairsden, CA). Washington, DC: American Forests.

Lee, Robert G. 1994. *Broken Trust, Broken Land: Freeing Ourselves from the War over the Environment*. Wilsonville, OR: BookPartners.

Lee, Robert G., Donald R. Field, and William R. Burch Jr. (eds.). 1990. *Community and Forestry: Continuities in the Sociology of Natural Resources*. Boulder, CO: Westview Press.

Lynch, Owen J., and Kirk Talbott. 1995. *Balancing Acts: Community-Based Forest Management and National Law in Asia and the Pacific*. Baltimore: World Resources Institute.

McCloskey, Michael. 1995. "Report of the Chairman of the Sierra Club to the Board of Directors." Unpublished report.

Mackie, Gerry. 1994. "Success and Failure in an American Workers' Cooperative Movement." *Politics and Society* 22(2): 215-35.

McLain, Rebecca, Margaret Shannon, and Chris Christensen. 1995. "The Forest Is More than Trees: Special Forest Products Politics in the Pacific Northwest." Paper presented at the Who Owns America? Conference, Land Tenure Center, University of Wisconsin, Madison.

McLain, Rebecca J., and Eric T. Jones. 1996. "Creating Space for Mobile Wild Mushroom Harvesters in Community-Based Forestry in the Pacific Northwest." Paper prepared for the International Association for the Study of Common Property Annual Meeting, 5-8 June, Berkeley.

Nelson, Robert H. 1995. *Public Lands and Private Rights: The Failure of Scientific Management*. Lanham, MD: Rowman and Littlefield.

Pavy, Gordon R. 1994. "Winning NLRB Elections and Establishing Collective Bargaining Relationships." In Sheldon Friedman et al. (eds.), *Restoring the Promise of American Labor Law*. 110-21. Ithaca, NY: ILR Press.

Peluso, Nancy L. 1992. *Rich Forests, Poor People: Resource Control and Resistance in Java*. Berkeley: University of California Press.

Plotkin, Sidney. 1987. *Keep Out: The Struggle for Land Use Control*. Berkeley: University of California Press.

Poffenberger, Mark, and Betsy McGean (eds.). 1996. *Village Voices, Forest Choices: Joint Forest Management in India*. Delhi: Oxford University Press.

Practitioner. 1996. A publication of the National Network of Forest Practitioners. Santa Fe, NM: Forest Trust.

Ratner, Shanna E. 1984. "Diversified Household Survival Strategies and Natural Resource Use in the Adirondacks: A Case Study of Crown Point, New York." MSc thesis, Cornell University, Ithaca, NY.

Reich, Charles. 1964. "The New Property." *Yale Law Journal* 73(5): 733-87.

—. 1990. "New Property after 25 Years." *University of San Francisco Law Review* 24(2): 223-71.

Richards, Rebecca T., and Max Creary. 1996. "Ethnic Diversity, Resource Attachment, and Ecosystem Management: Matsutake Mushroom Harvesting in the Klamath Bioregion." *Society and Natural Resources* 9: 359-74.

Ringgold, Paul, and Mary Mitsos. 1996. "Land Management Stewardship Contracts: Background and Legislative History." Paper presented at the Community Stewardship Meeting, sponsored by the Pinchot Institute for Conservation, Redding, CA, 3-4 October.

Robinson, Christina. 1994. "Multiple Perspectives: Rules Governing Special Forest Product Management in Coastal Washington." Master's thesis, College of Forest Resources, University of Washington, Seattle.

Rosenblum, Jonathan D. 1995. *Copper Crucible: How the Arizona Miners' Strike of 1983 Recast Labor-Management Relations in America*. Ithaca, NY: ILR Press.

Soukhaphonh, Savieng. 1997. "Non-Timber Forest Products, the Harvesters, Rules, and the Industry." Unpublished paper, Evergreen State College, Olympia, WA.

Tovares, Joseph. 1995. "Mojado Like Me." *Hispanic Magazine* May: 20-26.

United States Congress. 1998. Fiscal Year 1999 Omnibus Appropriations Bill. Section 347.

United States Forest Service, Kaibab National Forest, and Dixie National Forest. 1992. "Stewardship End Results Contracts (in Public Law 102-154)." Unpublished paper.

United States Senate. Senate Energy and Natural Resources Committee. Forest and Public Land Management Subcommittee. 1995. "Hearing Summary: Alternatives to Federal Land Management and Ownership." Unpublished paper.

Vaupel, Suzanne, and Philip Martin. 1986. *Activity and Regulation of Farm Labor Contractors*. Berkeley: University of California, Division of Agriculture and Natural Resources.

Wial, Howard. 1994. "New Bargaining Structures for New Forms of Business Organization." In Sheldon Friedman et al. (eds.), *Restoring the Promise of American Labor Law*. 303-13. Ithaca, NY: ILR Press.

Williams, P. (Representative). 1995. Forest Ecosystem Stewardship Demonstration Act of 1995 (H.R. 1682 Williams). United States House of Representatives, 104th Congress, 1st Session.

Yimsut, Ranachith. 1995a. *Understanding Asian and Southeast Asian-Americans: A Closer Look at the Social, Cultural, and Customs of Cambodian (Khmer), Lao, and Vietnamese-American*. Bend, OR: Deschutes National Forest.

—. 1995b. "Access to Our National Forests: Who Has the Rights? Issues and Concerns Facing the Southeast Asian Community in Oregon." *Practitioner* 4: 5-6.

Zekeri, Andrew. 1995. "The Racial Bifurcation of Community Development and Land Tenure: Implications for Community Development Practitioners." Paper presented at the Who Owns America? Conference, Land Tenure Center, University of Wisconsin, Madison.

Part 4: Policy Innovations

The war in the woods on the Pacific coast is unlikely to be definitively won or lost by any of the major combatants. Our best hope may be in incremental and local changes that create the conditions for truces. Chapters in Part 4 focus on such innovations in an effort to integrate new ways of thinking about forest stewardship. Clark Binkley proposes a broad land-use strategy to address a range of concerns about forest management in British Columbia. Neal Wilkins evaluates a discretionary policy approach to conserving habitat. Finally, Mae Burrows offers a political analysis of multistakeholder processes.

Clark Binkley offers land-use zoning, coupled with investment in intensive forest management, as a means to meet demands for wood fibre while protecting the myriad ecosystem functions performed by forests. This chapter was first published in *BC Studies* in 1997 and generated considerable response in the forest policy community. We republish it here because of the continuing attention given to the policy idea of zoning. Binkley argues that British Columbia lies at a crossroads between forests provided by providence and those created through human husbandry and stewardship. Large expanses of virgin forest remain in only a few places – in British Columbia and elsewhere in Canada, in eastern Russia, in the Amazon, and in parts of Africa. Unlike these developing nations, British Columbia has the wealth to treat its forests carefully. Moreover, unlike most other developed parts of the world, in this province there is still an opportunity to sustain a vast wild estate while continuing a prosperous society based on forest resources.

Binkley's chapter contemplates British Columbia's future by addressing three issues. First, Binkley identifies stages of resource depletion. Second, he defines some essential attributes of a sustainable forest sector. And third, he suggests how policies might adjust to move from the present position of seemingly intractable conflict to a preferable future. In an important sense, his proposal goes counter to the dominant ethos of this collection; it is a comprehensive plan framed by a global vision. But if zoning is conducted in a context of local and regional autonomy, then it may facilitate small-scale innovation.

Neal Wilkins reviews scientific, social, and economic barriers to effective conservation on private forest lands in the Pacific Northwest. He analyzes the impact of implementing the Endangered Species Act on private landowners' actions affecting management of fish and wildlife resources. Wilkins explores unintended consequences of various regulatory approaches and discusses how disincentives may be overcome using multispecies habitat conservation plans. Drawing on his experience as a wildlife biologist for Port Blakely Tree Farms, he analyzes these plans as a means to protect endangered species on private lands. He argues that, under particular conditions, these plans can serve the long-term interests of the public.

Both Binkley and Wilkins assume the necessity of trade-offs between industrial forestry and environmentalists. Both also maintain that the forest industry must embrace the righteousness of forest conservation and preservation and that environmentalists must accept the desirability of a robust, efficient forest products industry. Governments must respect the stewardship capacity of the private sector, and the private sector must respect the need for government regulation of the public goods produced by forests. Mae Burrows leaves these issues open as she addresses the role of political economic power in forest policy making. For Burrows, democracy rather than efficiency is the priority.

Burrows draws on her participant/observer study of the Commission on Resources and Environment (CORE) in British Columbia to offer generalizations about alternative dispute resolution processes. She identifies the essential components of an effective multistakeholder process and contrasts them with the CORE negotiations on Vancouver Island. Her discussion examines alternative dispute resolution, process design, agenda setting, participation models, funding problems, and definitions of consensus. While multistakeholder processes are ostensibly intended to represent the interests of all important sectors, Burrows argues that citizens not aligned with big timber, big environment, or big labour are at a disadvantage. However, instead of urging citizens to abandon such processes, she offers a strategy for their participation.

The chapters in this section vary considerably in perspective and focus. This variation facilitates consideration of pragmatic policy concerns (providing clear goals for land managers), concrete management imperatives (how to get companies to change their practices), and more abstract issues of democratic theory (how to ensure equal citizenship/participation in forest politics). Readers are unlikely to agree with all of the authors, but reflecting on the issues that they raise and the arguments that they make will contribute to productive participation in democratic conversation about forest management and use.

7

A Crossroad in the Forest: The Path to a Sustainable Forest Sector in British Columbia

Clark S. Binkley

> I shall be telling this with a sigh
> Somewhere ages and ages hence:
> Two roads diverged in a wood, and I –
> I took the one less traveled by,
> And that has made all the difference.
>
> – Robert Frost, "The Road Not Taken" (1916)

British Columbia lies at a crossroads between forests provided by providence and those created through human husbandry and stewardship. Many of the changes now confronting the province are predictable consequences of human interaction with primeval forests. Indeed, the earliest recorded story – the Epic of Gilgamesh written in cuneiform on a clay tablet 5,000 years ago – remarks on the dire consequences of forest depletion. Each subsequent civilization has relived this story with little change in the theme, from the Greeks in the Mediterranean to the wandering bands in central Europe, the Swedes late in the previous century, and our southern neighbours in the past decade or so.

Large expanses of virgin forest remain in only a few places – in British Columbia and elsewhere in Canada, in eastern Russia, in the Amazon, and in parts of Africa. Those in British Columbia lie on the cusp of an irreversible slide into the established historical pattern of resource depletion and attendant social disruption. However, unlike most other developed parts of the world, in this province there is still an opportunity to make the changes needed to sustain a vast wild estate while continuing a prosperous society based on forest resources.

This chapter contemplates British Columbia's predicament by responding to three questions. First, by way of background, what are the predictable stages of resource depletion as they have evolved in other parts of the world? Second, what are some of the key elements of a sustainable forest sector for the future? And third, how must policies adjust to move from the uncomfortable position of the present to a preferable future?

Timber Depletion and Forest Sector Development
To examine the predictable links between timber depletion, forest sector development, and societal evolution, I will use the economist's mode of analysis: hold everything else constant to focus clearly on the issue at hand, with full knowledge that the world contains a much richer fabric of complications. In this spirit, consider a sovereign entity (called a "country" below) embedded in a world forest sector. Imagine that this country discovers an old-growth timber resource large enough so that its development is significant in the relevant markets. In the absence of policy intervention, how does the forest sector develop?

The first part of this story is the comparatively old and well-known saga of the adjustment through time of a timber stock (Lyon 1981; Sedjo 1990; Sedjo and Lyon 1990). The second part links the dynamics of the timber stock to changes in the forest sector, macroeconomy, and environment more broadly (Vincent and Binkley 1992).

Dynamic Adjustment of the Timber Stock
In the early stages of development, net growth of the forest is nil; photosynthesis just balances the death of plant tissues and entire trees. Because growth is nil, any harvest exceeds the growth of the forest. Since harvest is greater than growth, the timber inventory declines.

As the inventory of old-growth timber declines, timber becomes more scarce. In economics, "scarcity" is a synonym for higher prices. Harvesting costs will increase as logging pushes into increasingly remote sites. Timber rents – the value of the standing timber itself – will increase as a consequence of old-growth depletion and the link between timber markets and capital markets. Prices rise until the purposeful husbandry of second-growth timber and the use of nonwood substitutes (stone, concrete, or steel for construction; fossil fuels, solar energy, and conservation for energy) become economic. Because timber prices have risen, other countries can compete in world markets, either by exploiting their own old-growth reserves (e.g., the Philippines, Indonesia, or Malaysia) or by developing plantations (e.g., New Zealand, Chile, Brazil, or South Africa).[1]

In the absence of subsidies, high-cost old growth is apt to remain unharvested and become economic wilderness.[2] We see this today in remote locations of British Columbia and other parts of the world. What do these adjustments in the timber stock imply for such important concerns as capital, labour, and the environment?

Capital and Labour

In economic terms, harvesting timber transforms ecological capital into economic capital. The reduction in timber stocks increases the supply of capital to the economy. Because of the increase in the supply of capital, its price will decline relative to the price of labour. Because timber prices are rising, the price of capital falls relative to the price of timber even more.

The technology used to process timber logically adapts to these changes in factor prices; firms substitute capital for labour and capital for timber. Because of the nature of the technologies used in the forest sector, the substitution of capital equipment for timber (e.g., investment in a sawmill that produces more lumber from a given input of logs) typically reduces specific labour utilization as well. The resulting increases in technical efficiency push up output per person hour (i.e., labour productivity) and open the door for higher wages. But the mathematical inverse of labour productivity is employment per unit of output, so gains in efficiency, all else being equal, mean fewer workers per unit of output. As a result of these effects, employment/cubic metre (cum) of timber harvested in British Columbia fell by a factor of two between 1961 and 1989 (Nixon 1991).

An iron law of economics holds that the economic return to the addition of any one factor of production declines as more of that factor is used. As firms use more and more capital in an effort to offset rising relative prices of timber and labour, the single-factor productivity of capital will decline. Lower returns on capital mean less capital investment. Unless technical efficiency increases via the application of new technologies, wage rates ultimately must fall (or at least not rise as much as they do in other countries), or the forest sector will fall into a self-perpetuating spiral of declining productivity.

Increases in timber prices will drive up product prices. Higher product prices encourage substitution away from conventional wood-based products. So, for example, the consumption of softwood lumber in the United States has remained roughly constant since 1900 at 40-50 billion board feet annually, despite a sevenfold increase in economic activity and more than a doubling of population in that country. Some of this substitution has been pure efficiency gain (e.g., because of improved knowledge about the performance of wood used in buildings, a basic framing member that measured 2" x 4" cross-section at the turn of the century has now been reduced to 1.5" x 3.5" with no loss of building safety), some has been wood-wood substitution (e.g., in sheathing applications, first plywood substituted for lumber; more recently, oriented strandboard has substituted for plywood), but some has replaced wood with other materials (steel studs, plastic bags, and concrete buildings).

Environmental Values

As per capita income increases, so too does the demand for the services of natural environments. Empirical evidence substantiates this claim for some features of the environment (e.g., air quality, water quality, and outdoor recreation), and the positive relation between income and environmental values is probably more broadly applicable as well.[3] At the same time, development of forests for timber production will decrease the supply of these environmental services.

Local and global environmental services operate outside formal markets, either because they are true public goods (e.g., aesthetically pleasing landscapes, carbon sinks) or because society has chosen not to allocate them through markets (clean water flowing from a forested watershed, recreation). It is a simple truism that such goods are systematically undervalued in forest consumption and production decisions. As a consequence, market-based patterns of forest use cannot and do not reflect the social values of these inputs and outputs. This mismatch between the social valuations of natural environments and their value in formal markets will widen over time. Once the gulf is sufficiently large to overcome transactions costs, institutions will emerge to place values on these services. They may actually be traded in formal markets, or they may be protected through direct government ownership, through regulation, or through elaborate and costly mechanisms of planning and public involvement in forest decision making. Whatever the mechanism, the costs of these formerly free environmental inputs will rise, and productivity as conventionally measured will decline as these previously free inputs come to carry positive costs.

Challenges for a Sustainable Forest Sector in British Columbia

The problems that British Columbia faces today are no more and no less than the local manifestation of the more general phenomena outlined above. In the absence of positive adaptations in public policy, the province faces wrenching structural adjustments in its economy, communities, and regional patterns of development. The well-known "fall down effect" in British Columbia – the planned reduction in timber harvests as old growth is depleted and investment in second growth is inadequate – is but one example.

In 1992, the BC government announced a review of long-term timber supply on all timber supply areas (TSAs) and tree farm licences (TFLs) – virtually the entire land base that supports industrial activities related to the province's forest sector. Under the Forest Act, the level of annual allowable cut (AAC) for individual areas is not calculated from any one formula but is set by the chief forester on the basis of broad biophysical and socioeconomic criteria. As a result, it is nearly impossible to predict in advance what future AACs might be. However, the analytical models used by the Ministry of Forests to estimate potential future harvest levels give an indication of the results of current policy direction. At this writing (1995), the ministry has released analytical reports for twenty-five of the thirty-six TSAs, representing some 35.3 mm cum of AAC (versus about 72 mm cum for TSAs and TFLs provincewide). These reports collectively suggest that current policy regimes will result in a long-term reduction of about 23.5 percent in provincial total harvest levels, with a significantly greater impact on the coast than in the interior (Miller, personal communication).[4]

What are the implications of this level of reduction in timber supply? A recent study (Binkley 1994) examined a variety of economic impact analyses related to harvest reductions. It concluded that a 25 percent reduction in harvest levels would mean a loss of up to 92,000 jobs and $4.9 billion in GDP in the province, with more than proportional impacts on government revenues (i.e., the net loss of taxes on GDP grossed up by increases in social service costs for unemployed workers). Although even Vancouver's economy relies heavily on the forest sector,[5] the impacts would be felt most strongly in the thirty-nine of fifty-five rural communities in the province where the forest sector is the dominant basic industry (Ministry of Finance and Corporate Relations 1992). These economic effects are likely to produce social effects that go along with community disruption (e.g., alcoholism, divorce, suicide). These social problems will come just when the province is least able to offer assistance; the study indicates that, with a 25 percent reduction in harvests, the provincial budget deficit will increase by about $2 billion.

The obviously large negative impacts of current policies suggest that significant benefits might accrue from a change in policy direction. We are indeed at a crossroads in the development of our forest sector. Before turning to desirable policy changes – the path best taken – let us examine some desirable policy outcomes. They include simultaneously maintaining harvest levels, enhancing the productivity of the forest sector at high wage rates within the constraints of available timber, and sustaining important ecological and environmental characteristics of British Columbia's diverse forest estate.

Maintaining Harvest Levels
Current policies are apt to lead to significant reductions in harvest levels. Are these reductions necessary?

Historically, British Columbia's forests have been managed extensively under the implicit assumption that virtually the whole forest land base would one day be available for timber production. The BC Forest Service and licensees incorporated nontimber values into the timber production plans through a process of "integrated resource management" that attempted to consider wildlife, riparian habitat, recreation, water flows, grazing, and other forest uses on each hectare where logging was to occur. Investments in silviculture were low. While licensees are now required to regenerate all areas logged to a "free to grow" stage, and massive reforestation efforts under the various federal/provincial agreements have virtually eliminated the backlog of "not satisfactorily restocked" (NSR) lands, British Columbia's use of silvicultural technology lags that in virtually every country with which the province competes.

This approach to land management has clearly failed. It does not satisfy those concerned with the nontimber values of the forests. It is a clear prescription for reduced harvest levels and does not respond to the commercial needs of the forest sector, the economic needs of communities, or the financial needs of the provincial government. Recent theoretical work (Swallow, Parks, and Wear 1990; Vincent and Binkley 1993) and empirical analysis (Sahajananthan 1994)[6] confirm the old idea that the multiple benefits of forests are best provided by zoning the forest into a series of special-use areas corresponding to the range of forest values that society demands – from wilderness to timber production. The vanguard of the environmental movement understands the wisdom of this approach (e.g., Alverson, Kuhlman, and Wallen 1994; McNeely 1993).

British Columbia moved tentatively in this direction through the process established by the Commission on Resources and the Environment (CORE). In the three areas studied (Vancouver Island, the Cariboo-Chilcotin, and the Kootenays), CORE defined zones ranging from protected areas to intensive timber production. The details of management in each zone remain to be developed; however, if the management rules actually permit intensive management on some of the land, then it may be possible to offset much of the planned reduction in harvest levels.

Consider a benchmark. In the late 1970s, Weyerhaeuser studied the biochemical efficiency of trees in turning sunlight into wood and modelled maximum biologically possible timber yields (Farnum, Timmis, and Kulp 1983). The company applied this model to two sites in the United States – one in the Pacific Northwest (Douglas-fir) and a second in the southeast (loblolly pine) – where it practises as intensive forestry as can be found anywhere in the world. For example, in the southeast, the study plantations were site prepared, bedded, fertilized, planted with genetically improved/mycorrhizal-inoculated seedlings, optimally spaced after planting, and repeatedly thinned and fertilized. Yet the production of these stands achieved only 40-50 percent of the theoretical yields. Natural stands in the same locations grow only about 10-25 percent of the theoretical yields.

Forests in British Columbia are managed much less intensively than are these study sites. As a consequence, there appears to be considerable latitude for increasing the production of economically usable plant parts in the province. With intensive silviculture, yields of from two to five times the levels attainable in natural stands generally appear to be economic.

There is some local empirical evidence that these kinds of increased yields are indeed feasible in British Columbia. Tolnai's (1991) analysis of Weyerhaeuser's TFL 35 near Kamloops found that, with more intensive management, harvest levels on this licence could be sustainably increased by 70.1 percent. In some stands, the increase was even more dramatic, rising from 2.3 cum/ha/yr under the current management regime to 8.3 cum/ha/yr in a more intensive management regime. And, remarkably, these management regimes involved only prompt restocking with desirable species (in this case, lodgepole pine) at an appropriate density. Increased yields from genetic improvement of planting stock and site emendations could be added to these gains.

The impediments to achieving these gains are primarily institutional. Licensees have no incentive to invest in more intensive silviculture because the gains accrue to the Crown and not to them. From a financial point of view, private investments in silviculture are equivalent to tearing up money and throwing it out the boardroom window; they neither enter the balance sheet nor produce incremental earnings in the future.

Enhancing Productivity

Economists define "productivity" simply as the value of an industry's outputs divided by the costs of its inputs. Increases in productivity are obviously prerequisite to increases in material standards of living. Porter (1990, iii) expresses the situation well: "Productivity is the prime determinant in the long run of a nation's standard of living, for it is the root cause of national per capita income." Porter goes on to observe that resource-based economies can continue to make productivity (and thus income) gains by improving production processes and products. Thus, natural resource industries are no less desirable than high-tech industries.

Productivity in the BC forest sector is squeezed between a rising floor of raw material costs and a fixed ceiling for product prices set by international competitors in the forest products industry and by the cost of substitute products. As a result, the BC forest sector faces enormous challenges in maintaining the high levels of productivity[7] that have produced the high standard of living enjoyed by the province, let alone the high levels of productivity growth needed to sustain this standard of living in the face of the predictable negative changes pressing the sector.

Historical reliance on old-growth timber that required no human inputs to grow meant that the value of timber – the rent – was available for distribution to whatever parties found political favour. Historically, some of the rent flowed to licensees as an inducement to establish processing facilities.[8] Some flowed to labour in the form of higher wage rates.

To counteract the negative competitive effects associated with rising timber costs and high wage rates, British Columbia needs higher than average productivity growth. Productivity growth can occur through reductions in the total costs of inputs or through increases in the values of outputs. As we have seen, there is endogenous upward pressure on raw material costs, and reductions in our high wage rates are not socially attractive. As a consequence, the only acceptable, sustainable answers seem to lie in more efficient production (Binkley 1994).

Increases in productivity result from investments in technology that are either earlier or better than those made by competitors. Yet in British Columbia the forest sector spends only about 0.7 percent of gross receipts on R&D, whereas Sweden, for example, spends about 1.8 percent (Binkley and Watts 1992). Despite these problems, some elements of the forest sector have been able to adopt technology rapidly. However, it is not clear that this increase in technical efficiency was adequate to offset higher production costs, competitors' responses, and marketplace effects (Binkley 1994).

Sustaining Environmental Quality

Imagining that British Columbia devotes some 20 million hectares to zones where timber production is a significant or dominant use, then a vast area remains to support other forest uses. Urban and agricultural lands comprise a fairly small portion of the province, so perhaps 65 million hectares could be devoted, more or less exclusively, to sustaining the environmental values that forests provide. This area would include at least 23 million hectares of productive forest, 20 million hectares of savannahs and other vegetated land, and 20 million hectares of "rocks and ice." As a point of comparison, the area of France equals about 55 million hectares. In other words, through a policy of zoning and intensive, dominant-use management, British Columbia could devote an area larger than France to sustaining nontimber values of the forest.

Such a policy might involve three zones: intensive timber production areas as described above, strictly protected areas, and integrated management joint-use areas to form the transition between the other two landscapes. Making such a strategy work requires, first, that a core of protected areas is selected in a way that includes reasonable representation of the great variety of BC ecosystems in large enough blocks to sustain landscape-scale processes. The current provincial Protected Areas Strategy is well suited to this task. Second, the joint-use areas must be managed in a way that responds to environmental needs. This will occur nominally through the Forest Practices Code, but there may be serious environmental problems associated with the approaches that it specifies. The code will reduce the average size of clear-cuts and will require that areas logged "green up" (i.e., reach a specified age or cover condition) before adjacent areas may be logged. While sounding innocuous, if not beneficial, the net effect of these two provisions is to scatter the harvest across the landscape. This pattern of harvests will fragment the forest with predictable consequences for biological diversity (Harris 1984). The area of forest edge habitat will increase, and the area of forest interior habitat will decrease. These landscape changes will tend to favour early successional species, while much contemporary public concern focuses on late successional ones (e.g., the spotted owl).

The Forest Practices Code will also require more rapid development of the road system and maintenance of a greater amount of it each year. The greater length of an active road network will provide more opportunities for illegal hunting. By building the road system faster, British Columbia will foreclose in the future the option of deciding that an area should remain roadless. Since the worst environmental problems associated with timber production in the province come from the failure of roads, it is ironic that new legislation purporting to protect environmental values actually demands more road building.

Finally, the intensive management areas must be managed very intensively for timber production to make up the harvests lost through the creation of protected areas and low-intensity management areas. The Forest Practices Code logically should contain a different set of standards for these intensive management areas than for the other zones. Control of the collateral environmental damage frequently associated with logging should occur through land-use designations and not through management regulations on the intensive management areas. Policy-driven constraints such as minimum rotation ages, green-up/adjacency, and visual quality objectives should be relaxed in the intensive management areas. The tenure system should provide stronger incentives for investment in timber production on these areas. Applied research should be focused on these areas to adapt the lessons on intensive timber management from locations such as Brazil, New Zealand, the United States, and Sweden to the specific circumstances found in British Columbia.

Needed Policy Adjustments

Despite the vast size of the province and its relatively small population, land has become scarce in British Columbia. Each hectare seems to face multiple demands – from local residents, from the province as a whole, and from the international community. The increased scarcity of land logically implies that other productive factors – knowledge, capital, and labour – should be substituted for land to produce desired outcomes, whether environmental values or timber production. Government ownership and control of land in British Columbia means that the price system has not been able to signal the needed changes. Like a fault line that has accumulated strain over years of tectonic action, the BC forest sector is close to rupturing. Abrupt change appears to be inevitable. Good policy would help to ease the transition by providing flexibility for change. What are some appropriate actions?

Reform of Forest Institutions

A thin membrane of institutional arrangements mediates the interactions between humans and forests. In British Columbia, the most significant institution affecting forests is the tenure system. The basic concepts used in the current system of forest tenures derive from the 1945 royal commission (the Sloan Commission). At the time, the basic societal need was to use old-growth forests as a wooden magnet to attract capital investment in the processing sector. Capital investment meant economic development, particularly in the hinterlands of the province. Environmental values of the forest were not of great concern, in part because so vast an area of forests remained undeveloped.

Times have changed. Societal needs now involve attracting capital to invest in the forests themselves, in activities that produce value-added products, and in activities that use secondary sources of fibre such as bark, sawdust, planer shavings, and trim blocks. Standing timber is not a particularly logical or effective inducement for these kinds of investments, so the current tenure system is a blunt tool for crafting the future. As a result of increases in the timber fees collected by the provincial government and increasingly restrictive operating rules, forest tenures are no longer the potent plums of political advantage that they once were.

The creation of Forest Renewal BC (FRBC) nominally provides the capital needed for investments in forest management. Unlike the many previous "permanent" silvicultural funds,[9] increased stumpage fees and royalty payments directly fund FRBC without reference to annual appropriations from general revenues. An independent board of directors will direct these funds to five areas: silvicultural investments, industry diversification and value-added manufacturing, environmental restoration, strengthening communities, and training workers, with R&D being directed at all these areas. Although there are similar arrangements elsewhere (e.g., the K-V funds for national forest lands in the United States), the magnitude of the FRBC fund is unique. Planning for FRBC assumed that the bellwether grade of lumber (interior spruce-pine-fir, called by the trade acronym SPF below) would trade at $US 350/mbf – at this price, the pool of available funds would be about $400 million per year.

While preferable to the previous situation, in which little was invested in growing trees, the FRBC approach suffers from four serious shortcomings. First, the anticipated revenues may not be realized. SPF traded at around $300/mbf at the end of 1994, and respected industry analysts forecast even lower prices for 1996. Although historically low North American interest rates and restrictions on Canadian-US lumber trade forced prices up well beyond the anticipated levels, these prices probably cannot be sustained in the face of pressure from substitute products. At price levels below $US 300/mbf, the size of the fund is quite sensitive to fluctuations in product prices; each $1 reduction in product price results in a $0.58 product-equivalent reduction in the stumpage rate (Binkley and Zhang 1994). Falling product prices will jeopardize the entire program, and it is not at all clear that investments in the land will be able to secure priority allocation of whatever funds are available.

Second, the board is not constrained to invest in the best silvicultural opportunities; indeed, subsection 6(b) of the BC Forest Renewal Act *requires* the board to "provide advice to Forest Renewal BC as to appropriate *regional* goals for expenditures" (emphasis added). The board is contemplating devolution of funding authority to regional bodies. In the absence of clear guidance from the act, the regional allocations are apt to be governed more by politics than by investment efficiency.

Third, FRBC has no particular incentive to be frugal in how it administers its funds. For example, an early submission to the board from the Ministry of Forests and the Ministry of Environment, Lands and Parks requested $33.8 million per year to add perhaps 400 full-time equivalents in the two bureaucracies (Zak, personal communication).

And fourth, the efficacy of silvicultural investments depends on an intimate knowledge of the land base, local ecological conditions, and overall landscape management objectives. It is doubtful that a distant, third-party funding agency will have adequate understanding of these local circumstances to invest wisely.

Discussions in British Columbia about forest tenures frequently use the analogy of owning a house outright (private land) versus renting it (tree farm licences or forest licences). If you lease a house to someone, you hardly expect that person to paint the house (to invest in silviculture), and if you require that person to do so (free-to-grow requirements) he or she will probably use the cheapest paint possible. To extend this analogy, FRBC offers to buy the paint but provides no incentive to apply it carefully (to implement the most efficient silvicultural investments).

At the same time, there is ample empirical evidence – both from elsewhere in the world and now from British Columbia – that strengthened property rights lead to higher levels of private investment in forest management.[10] In British Columbia, stronger property rights could be achieved either through outright privatization of those lands in the intensive management zones created by CORE or through the sale of long-term leases along the lines of New Zealand's recent policy. Existing tenure holders might retain their current lands or be satisfied simply to buy logs from others. New kinds of organizations would probably enter the field. For example, pension funds now own timberland in the United States, New Zealand, and Chile worth more than $4 billion. The largest of the timberland management organizations serving pension funds – the Hancock Timber Resource Group – now owns and manages about 30,000 hectares in British Columbia. Stronger private property rights would not only increase the pool of capital available to invest in forest management but also provide greater flexibility in responding to the rapidly changing circumstances now found in the province.

Tenure reform should be accompanied by reform of other dysfunctional policies related to the tenure system. Perhaps the most significant of these are the appurtenancy clauses in many coastal tenures that, on penalty of licence forfeiture, constrain logs from a specific licence to flow to a specific mill. The original objective of these clauses was to use timber to support the development of individual communities. But they have become very expensive subsidies for regional development. In times of tight log markets, efficient mills not associated with the restricted licences must close, while the less efficient ones that are appurtenant to a particular licence continue to operate. The difference in log values between the efficient and inefficient mills is significant – in one case, it amounts to about $60/cum. In this case, the subsidy for each employee of the appurtenant mill is about $75,000 per year in addition to his or her wages. I suspect that the employees involved would be pleased to receive a fraction of the capitalized value of this amount in return for agreeing to leave their jobs without complaint to the government.

Strengthening Transitional Mechanisms
The policy changes required to put the BC forest sector onto a path of economic, social, and environmental sustainability will inevitably produce losers as well as winners. Insofar as those old policies protect the self-interests of potential losers, it is natural and predictable for them to hold fast to existing policies and to block the needed adjustments. The magnitude of the adjustments needed in the forest sector demands that explicit attention be paid to the problems of policy transition (Behn 1978).

If policy changes improve economic efficiency (and many in the province would), then – by definition – the winners would be able to compensate the losers and still be better off than they were before the policy change. As a practical matter, those who wish to see policy changes – and will benefit from them – should support compensation to those who lose as a result of such changes.

From a more philosophical vantage point, British Columbia has explicitly chosen to allocate timber resources not through market mechanisms but through a process of government control. Timber prices do not properly signal the relative scarcity of timber, and firms' harvesting decisions are virtually dictated by government representing society as a whole. Firms deploy capital, and labour moves to specific regions as a result of these socially determined decisions. Therefore, society as a whole logically bears the burden of unanticipated and unannounced adjustments to the system.

Compensation should be paid to those who are economically injured by reductions in timber harvests, changes in tenure policies, or other changes in forest policy. Potentially injured parties include those who own capital or labour made obsolete by the policy change. Compensation should cover, for example, the reduced values of homes, the costs of necessary job retraining, the lost values of tenures, and the forgone returns to productive capital such as sawmills and pulp mills made redundant by the policy change. Compensation for lost tenures might involve granting stronger property rights to a licensee over a smaller portion of a current licence. Compensation is best paid in kind, so timber somewhere else (perhaps created by increased silvicultural investments) can compensate for timber lost in one location, or another kind of job of comparable worth can compensate for a job lost in logging or sawmilling.

Creating a Knowledge-Based Forest Sector

Rapid adoption of improved technology is key both to international competitiveness of the BC forest sector and to responsible stewardship of the environment (Binkley 1994). Forest sector R&D expenditures in British Columbia are small. A significant gap with the province's competitors exists both for forest-related R&D and for forest-products R&D (Binkley and Watts 1992). Yet rapid development and adoption of leading-edge technology is a fundamental element of the path to a sustainable future.

BC firms face three daunting tasks in effectively deploying R&D (Binkley 1994). First, commodity-grade products (softwood construction lumber, market pulp, newsprint) dominate the BC industry. R&D leverage is less for commodities than it is for higher value-added products. By definition, commodities compete on the basis of production costs. Cost-reducing technology is likely to be available to all producers, so it creates no unique competitive advantage once all producers adopt it. As a consequence, only a strategy of rapid adoption will produce competitive advantage. Such a strategy is extremely difficult to implement. In this context, R&D may be better viewed as a powerful means of transforming the strategic direction of a company out of commodity businesses. MacMillan Bloedel's creation of Parallam® and Space Kraft® and its Nexgen coated paper project are examples of using R&D to move to value-added products.

Second, British Columbia is a leading international exporter with a large share of many of the markets that it serves. This market position exacerbates the problems of being a commodity producer. In these circumstances, some of the benefits of cost-saving technology are simply passed on to consumers. As long as the consumers are in one's own country, there is a case for government support of R&D, but when the customers reside elsewhere such an argument obviously does not apply. In these circumstances, effective technology strategies again involve rapid adoption and exploitation of features unique to the province (e.g., western red cedar).

Third, as a result of the tenure system, BC firms cannot exploit synergies between the design of forest management regimes and the design of new products and processes. Through its silvicultural regulations, the BC Ministry of Forests specifies the characteristics of future timber supply for the entire province (species mix, genotype, diameter, clear length), and all producers must adapt to these constraints regardless of their own market information. Under current institutional arrangements, the kind of R&D that led to the clearwood regime for *Pinus radiata* in New Zealand – an approach that permits quickly grown second-growth timber to substitute in many uses for British Columbia's old-growth *Pinus ponderosa* – would not be possible. Similarly, exceedingly low wood costs for Brazil's Aracruz pulp mill came from heavily targeted R&D to produce, plant, and process *Eucalyptus spp.* clones of high cellulose content. Such a strategy is unavailable to BC firms operating on public lands.

A high-technology strategy for the forest sector will create benefits beyond those associated with the forest sector alone. The sector has some strong backward links with the high-technology sector that could be strengthened and exploited more effectively. Vancouver is the heart of an important international forestry services industry. The Lower Mainland is a hotbed of technology firms that provide log- and lumber-scanning equipment and real-time sawmill optimization software. One equipment producer on Vancouver Island is among the world's leaders in the design and manufacture of cable logging equipment. Yet the provincial and federal governments have not, as a matter of policy, worked to build on these strengths.

Finally, a technology-based strategy will pay significant benefits for environmental quality. Better manufacturing efficiency is a powerful lever for environmental improvement. For example, from the perspective of product markets, even a modest gain in technical efficiency would offset reductions in timber harvests associated with the decision to set aside the Kitlope – a 320,000 hectare drainage on the north coast thought to be the largest unlogged temperate rainforest watershed anywhere in the world. As another example, progressively more sophisticated use of wood wastes has permitted substantial improvements in air quality in towns where sawmills operate. In the mid-1950s, all slabs, edgings, planer shavings, and sawdust (which collectively comprise about half of any log) for a typical interior sawmill were burned as wastes. Then pulp mills operating on wood residues were established, leaving only the sawdust and planer shavings to be burned. Now technology has advanced further to permit the production of medium-density fibreboard (MDF) – a high-quality panel used in furniture and similar end uses – from these residues. Removing all of the wood fibre from the waste stream leaves only the bark, and mills burn much of it for process heat used in the manufacture of lumber, MDF, and pulp.

Improved silvicultural technology – from better inventory and yield information to sophisticated techniques of molecular genetics – can sustain harvest levels on a smaller land base, freeing land for other uses. The power of this technology has not been extensively used in British Columbia, but it has been in other parts of the world. For example, because of an aggressive, high-technology plantation program, forest companies in New Zealand no longer log in that country's native forests but rely entirely on plantation forests. Their agreement to refrain from logging in natural forests – the Tasman Forest Accord – had virtually no economic impact on the country. In contrast, such an agreement in British Columbia would close over 90 percent of the forest industry, largely because the province has made no similar investments in R&D and forest management.

Conclusions

In husbanding its forests, British Columbia faces an ancient challenge. Paths travelled by earlier societies have led to unpleasant destinations. Past policy and development in the province have brought it to a crossroads in the forest. Only "the one less traveled by" will create a sustainable forest-based economy for the future while maintaining the renowned ecological and environmental features of British Columbia's magnificent forested landscape.

Following this path will require a massive redirection of current policies, both public and private. Many policies that have served well in the past are dysfunctional guides to the future.

Some of the needed changes are now under way. Land-use planning through the CORE process will provide greater long-term political certainty in the forest sector. Increased certainty is prerequisite to the high level of capital investment – in forests and in new, sophisticated processing equipment that sustainability, in its broadest sense, requires. The Forest Practices Code will provide a framework for guiding management in the different land-use zones. FRBC may be able to provide the capital required to finance this transition.

But these policy changes must be carefully implemented and strongly reenforced if they are to be successful. Once land-use zones have been established, various interests will no doubt seek to poach across the boundaries. The government must wisely distinguish legitimate needs to revise land-use zones from simple rent seeking. Economic instruments such as those increasingly used for pollution abatement may be helpful in drawing these distinctions.

The management rules for the various zones must be carefully crafted to support the distinct management objectives of each zone. Just as industrial intrusion on protected areas should be strictly limited, regulatory intrusion on intensive management areas should be carefully proscribed. Differences in Forest Practices Code regulations among the various zones provide a useful measure of success in this regard.

Once land-use zones have been defined and zone-specific Forest Practices Code rules have been written, it will be possible to craft a set of institutional arrangements and land tenures that more productively serves the sector. These arrangements logically involve significant public control in zones where public values dominate and significant private control in zones where private values are most prominent. In the case of parks and protected areas, institutional reform will require a much stronger parks agency to handle the capital investment and management activities needed to ensure that the protected areas and very low intensity zones in fact provide the environmental values anticipated from them. In the case of integrated resource management zones, institutional reform will require either stronger direct public management activities or a more sophisticated set of licence documents than is currently used. Economic instruments that bring licensee and public interests into line merit careful attention (e.g., pricing systems for recreation, water flows, or site degradation). In the case of intensive timber management areas, tenure reform will require more powerful inducements to make highly effective investments in timber production. Ample evidence from British Columbia and elsewhere suggests that markets operating through ordinary private property rights provide the needed incentives. Experiences from elsewhere – especially from New Zealand, Sweden, and the United States – will provide useful guidance on the advantages of different kinds of private ownership and of different mixes among the kinds of private ownership (small and large; institutional, industrial, and individual), but the unique circumstances in British Columbia will no doubt require unique approaches to strengthening private property rights in forest land.

Change is always uncomfortable, and, when the stakes are as high as they are in the BC forest sector, discomfort invites paralysis. Sensitive attention to the problems of transition can reduce the discomfort and invite more rapid, creative, and positive responses from the parties involved. Principles of compensation should be articulated at the outset. New tenure arrangements should be made sufficiently attractive so that at least some licensees will voluntarily adopt them. Because no one really knows the optimal approach for a sustainable future, experimentation (and its concomitant, failure) should be encouraged. Different approaches might suit different areas.

Progress down the new road – the one that leads toward a sustainable future – will require, most of all, new ways of thinking about old problems. The forest industry must embrace the righteousness of forest conservation and preservation, and environmentalists must accept the desirability of a robust, efficient forest products industry. Governments must respect the stewardship capacity of the private sector, and the private sector must respect the need for government regulation of the public goods produced by forests. This revolution of the mind will not be easy, but it is the only means down the path less travelled by.

Notes

This chapter is reprinted with the permission of *BC Studies*. An earlier version appeared in *BC Studies* 113 (Spring 1997): 39-61.

1   In broad lines, this is consistent with the development of the forest sector in the United States (Clawson 1979; Sedjo 1990). Harvest exceeded growth until the 1950s. Increased scarcity drove up timber prices, and high prices forestalled the predicted timber famine by choking off demand for wood products and encouraging investments in forest management. Timber prices rose at a real rate of about 4.6 percent per year between 1910 and World War II and about 3.1 percent per year from that period to the mid-1980s (Binkley and Vincent 1988).

2   This assumes that the cost of harvesting the last old growth is greater than its value once harvested. See Clark (1973) or Page (1977) for a discussion of the economics of extinction.

3   This relationship between income and environmental values evidently is not new (Perlin 1991: 120). Seneca best articulated the romantic view of forests shared by many of the leisure class of his time: "If you ever have come upon a grove that is full of ancient trees which have grown to unusual height, shutting out the view of the sky by a veil of pleated and intertwining branches, then the loftiness of the forest, the seclusion of the spot and the thick, unbroken shade on the midst of open space will prove to you the presence of God." One cannot help but note the similarity of this comment, made nearly two millennia ago, to contemporary descriptions of old-growth forests in British Columbia.

4   My analysis of twenty-two of the thirty-six reports suggested a twenty-year reduction in AAC in the interior of 11.8 percent, on the coast of 25.1 percent, for a reduction in the provincial total AAC of 15.9 percent.

5   Park (1991) estimated the economic impact of the BC forest industry in the metropolitan area of Vancouver for 1989 to be $6 billion of GDP and 115,000 jobs with wages and salaries of $3 billion. A more recent study (Chancellor Partners 1994) found that in 1993 133,000 jobs (one in six) and $6.2 billion of regional GDP in metropolitan Vancouver depended on the forest sector.

6   This study of the Revelstoke Forest District found that, through moderately increased management intensity, about 40 percent of the land base would produce the same amount of timber as would 100 percent of the land base under current rules for integrated resource management.

7   At present, BC mills are, on average, mediocre competitors in newsprint and pulp but – at least in the interior – are internationally competitive in lumber (NLK 1992; Simons 1992).

8   Before the April 1994 increase in stumpage rates, analysts commonly estimated the uncollected rents at about $10/cum. The increased stumpage payments announced in April averaged about $11/cum, so the amount of uncollected rent, especially for licensees with marginal mills, is now probably small. This is roughly consistent with the estimates of Binkley and Zhang (1994), which found that the change in the stumpage system

reduced the capital value of publicly traded firms in British Columbia by about $1.3 billion or about $2.8 billion if grossed up to the sector as a whole.

9 Since the original draft of this chapter was prepared in 1995, the recently elected NDP government announced its intention to take $400 million from FRBC's funds to offset the provincial government's operating deficit.

10 A study by Zhang (1994) confirmed empirically this well-known theoretical argument. He examined expenditures on silvicultural activities on different kinds of tenures while holding a variety of factors that might affect such investments – such as biogeoclimatic zone, site quality, or location – constant. After controlling for all of these factors, silvicultural investments are strongly correlated with the strength of property rights. Taking the weakest form of tenure – forest licences – as the base, expenditures per hectare on tree farm licences are 27.4 percent greater, and expenditures per hectare on private land are 81 percent greater.

References

Alverson, W.S., W. Kuhlman, and D.M. Wallen. 1994. *Wild Forests: Conservation Biology and Public Policy*. Washington, DC: Island Press.

Behn, R.D. 1978. "How to Terminate a Public Policy: A Dozen Hints for a Would-Be Terminator." *Policy Analysis* 4(3): 393-413.

Binkley, C.S. 1994. "Designing an Effective Forest Sector Research Strategy for Canada." Unpublished paper, Faculty of Forestry, University of British Columbia, Vancouver, BC.

Binkley, C.S., and J.R. Vincent. 1988. "Timber Prices in the US South: Past Trends and Outlook for the Future." *Southern Journal of Applied Forestry* 12: 15-18.

Binkley, C.S., and S.B. Watts. 1992. "The Status of Forestry Research in British Columbia." *The Forestry Chronicle* G8: 730-35.

Binkley, C.S., and D. Zhang. 1994. "The Impact of Timber-Fee Increases on BC Forest Products Companies." Unpublished manuscript, Faculty of Forestry, University of British Columbia, Vancouver, BC.

Chancellor Partners. 1994. "The Economic Impact of the Forest Industry on Metropolitan Vancouver." Study prepared for the Vancouver Board of Trade.

Clark, C.W. 1973. "Profit Maximization and the Extinction of Animal Species." *Journal of Political Economics* 81: 950-61.

Clawson, M. 1979. "Forests in the Long Sweep of American History." *Science* 204: 1168-74.

Farnum, P., R. Timmis, and J.L. Kulp. 1983. "The Biotechnology of Forest Yield." *Science* 219: 694-702.

Harris, L.D. 1984. *The Fragmented Forest*. Chicago: University of Chicago Press.

Lyon, K.S. 1981. "Mining of the Forest and the Time Path of the Price of Timber." *Journal of Environmental Economics and Management* 8: 330-44.

McNeely, J.A. 1993. "Lessons from the Past: Forests and Biodiversity." Unpublished paper, IUCN, Gland, Switzerland.

Ministry of Finance and Corporate Relations. 1992. "British Columbia Community Dependencies." Paper prepared by the Planning and Statistics Division for the Forest Resources Commission.

Nixon, R. 1991. "Comparative Data Charts Explain Forest Management Policies." *Forest Planning Canada* 7: 32-45.

NLK. 1992. "The Pulp and Paper Sector in British Columbia." Discussion paper prepared for the Forest Summit Conference, Vancouver.

Page, T. 1977. *Conservation and Economic Efficiency*. Baltimore: Johns Hopkins University Press.

Park, D.E. 1991. *The Forest Industry's Role in Vancouver's Economy*. Vancouver: Vancouver Board of Trade.

Perlin, J. 1991. *A Forest Journey: The Role of Wood in the Development of Civilization*. Cambridge, MA: Harvard University Press.

Porter, M. 1990. *The Competitive Advantage of Nations*. New York: Free Press.

Sahajananthan, S. 1994. "Single and Multiple Use of Forest Lands in British Columbia: The Case of the Revelstoke Forest District." Report submitted to BC Ministry of Forests, Revelstoke Forest District.

Sedjo, R. 1990. "The Nation's Forest Resources." Discussion Paper ENR 90-07, Resources for the Future, Washington, DC.

Sedjo, R.A., and K.S. Lyon. 1990. *The Long-Term Adequacy of World Timber Supply*. Baltimore: Johns Hopkins University Press.

Simons, H.A. 1992. "The Wood Products Sector in British Columbia." Discussion paper prepared for the Forest Summit Conference, Vancouver.

Swallow, S.K., P.J. Parks, and D.N. Wear. 1990. "Policy Relevant Nonconvexities in the Production of Multiple Forest Benefits." *Journal of Environmental Economics and Management* 19: 264-80.

Tolnai, S. 1991. "Addition Value to Our Heritage through Silviculture." Paper presented to the Western Silvicultural Contractors Association, 5 February, Vancouver.

Vincent, J.R., and C.S. Binkley. 1992. "Forest-Based Industrialization: A Dynamic Perspective." In N.P. Sharma (ed.), *Managing the World's Forests*. Dubuque, IA: Kendall-Hunt, for the World Bank.

—. 1993. "Efficient Multiple-Use Forestry May Require Landuse Specialization." *Land Economics* 69: 370-76.

Zhang, D. 1994. "Implications of Tenure for Forest Land Value and Management in British Columbia." PhD diss., Faculty of Forestry, University of British Columbia, Vancouver.

8

Wildlife Conservation on Private Lands: Habitat Planning and Regulatory Certainty

R. Neal Wilkins

While recent gains in the science of wildlife ecology are significant and of increasing sophistication, progress in applying this ecological knowledge to wildlife conservation cannot be similarly demonstrated. Perhaps this is because effective wildlife conservation programs require a combination of applied ecological knowledge and effective policy process (Mangel et al. 1996). Recent experiences in North America have demonstrated that poor process rather than poor knowledge is to blame for numerous resource management failures (Hilborn 1996). The track record of wildlife conservation efforts in the United States – particularly on private lands – provides numerous examples of this failed conservation effort (Wilcove et al. 1996). Although an understanding of ecological processes underlying wildlife conservation is largely gained through scientific experimentation, insight into organized implementation of wildlife policy is often obtained only through the crucible of application and controversy. The successes and failures of these policies – as they are implemented – teach us what works and what does not. Perhaps the most challenging arena for implementing effective conservation programs is on private lands – especially lands that are intensively managed but supply suitable habitat for rare, threatened, or endangered species.

Although wildlife conservation on private lands might be an obvious public benefit, the incremental land management decisions most important in realizing this benefit are made by individuals constrained by the need to realize profits and demonstrate long-term financial gains. For these individual landowners to consistently forgo financial and other self-interests to assure a greater public benefit would be inconsistent with basic free-market behaviour (Anderson and Leal 1991). In many cases, centralized regulatory processes require landowners to do just that. But landowners do not always comply with these requirements. The consequences of their behaviour may be outcomes that are unintended and often contrary to the goal of conserving important fish and wildlife habitats. These unintended consequences have limited the successful implementation of state and federal conservation programs aimed at recovering threatened and endangered species residing on private lands.

The federal Endangered Species Act of 1973 (ESA) is regarded as the strongest existing tool in the United States for protecting biological diversity (Patlis 1996). The stated purpose of ESA is to provide a means for conserving threatened and endangered species and the ecosystems on which they depend – basically, to protect biological diversity. Through ESA, Congress has established its intent to temper economic development in order to prevent further extinction of plants and animals. On private lands, the policy and legal controversy has focused on whether or not regulation of land use for the purpose of conserving endangered species constitutes a *taking* of private property for a public good – thus requiring compensation under the Fifth Amendment to the US Constitution. This contrasts with an alternative viewpoint that ESA requires the government to protect against harm, thus allowing prohibition of harmful actions without compensation (Rolston 1991). Regardless of the merits of each argument, private landowners and government regulators are confronted with the practical realities of meeting the current mandates of ESA while assuring continued commercial land use. This challenge requires use of regulatory processes that effectively incorporate the ecological sciences with operational considerations of commercial resource management.

Although scientific theory and applications for conserving endangered species have progressed substantially since ESA was passed, conservation efforts carried out under ESA authority have not provided for fundamental elements of recovery for most endangered species, especially those that depend on private lands (Wilcove et al. 1996). This is a generalization, of course, because there are several success stories on both public and private lands.

To assure that successful conservation efforts on private lands become the rule rather than the exception, planning processes must change at both federal and state levels to capitalize on the self-interests of private landowners. This must be accomplished by decentralizing protection and management decisions and placing them in the hands of management teams that include both regulators and local land managers, thereby forcing trade-offs that simultaneously meet the needs of landowners and wildlife conservation. I focus on recent examples from private forest lands in the US Pacific Northwest, discussing some unintended consequences of various regulatory processes and how disincentives may be overcome with collaborative planning efforts such as multispecies habitat conservation plans.

ESA Process Problems on Private Lands

Who Enforces ESA?

ESA is officially administered by federal agencies, but state agencies also play central roles in endangered species conservation. Federal species listings and guidance are the driving force behind regulations adopted by state agencies. This is the case in both Oregon and Washington. For example, 1990 guidelines proposed by the US Fish and Wildlife Service (USFWS) for avoiding the incidental take of northern spotted owls (*Strix occidentalis*) were subsequently adopted by the Washington State Forest Practices Board and enforced by the Washington State Department of Natural Resources. Likewise, under pressure from the National Marine Fisheries Service (NMFS), Oregon has initiated changes in the forest practice regulations to mitigate the risk of additional ESA listings for various stocks of Pacific salmon along the Oregon coast. These actions cause confusion among some landowners in distinguishing between federal and state regulations.

Do Incidental Take Prohibitions Work?

With respect to ESA, the federal authority to regulate actions on private lands lies primarily in the incidental take prohibition of Section 9. As defined in the act, "take" means to "harass, harm, pursue, hunt, shoot, wound, kill, trap, capture, or collect, or attempt to engage in any such conduct." "Harm" is defined in federal regulations to include "significant habitat modification or degradation where it actually kills or injures wildlife by significantly impairing essential behavioral patterns, including breeding, feeding or sheltering" (50 CFR 17.3). This definition of harm was the subject of the Supreme Court's 1995 *Sweet Home* decision, in which the court found in favour of the federal government, thus affirming the definition of harm in the ESA regulations.[1] From a practical perspective, even following the *Sweet Home* decision, the legal definition of harm remains ambiguous. Although legal scholars continue to debate the definition of harm, the idea that habitat modification could result in incidental take seems to prevail among many field biologists and agency practitioners – and this is the level at which most land managers must operate.

Conservation efforts on private lands are at times hindered by unintended consequences of implementing incidental take prohibitions – especially those that require site-specific protection at known occurrences. This is a complex problem without a simple solution, but some general pitfalls become apparent from implementation efforts.

Incidental take prohibitions on private lands are difficult to enforce due to inherent variability of biological systems. Prohibitions on private forest lands include adverse habitat modification caused by road construction or timber harvesting within occupied suitable habitats of listed species. This is a difficult prohibition to enforce because of individual circumstances of time and place (i.e., natural variability). Habitat modification that is genuinely adverse for one species may not affect another. Some species, for example, are limited by inadequate foraging resources in a portion of their range and inadequate breeding habitats in other portions. Likewise, among habitats considered suitable for a species, there remains variability in productivity. Absolute habitat thresholds for an area needed to support a population or breeding pair cannot be reliably determined. However, to accommodate enforcement efforts, incidental take guidelines are generally constructed with absolute thresholds. These enforcement tools are often derived from sparse and highly variable data sets that require some subjectivity in interpretation. Because they are to be applied broadly, the interpretations can overlook large differences in biological potentials among localized habitats.

Implementing incidental take prohibitions on private lands often favours short-term protection of individual animals at the expense of long-term conservation of populations. The definition of incidental take includes harm to individual animals through habitat modification. ESA regulatory efforts on private lands have therefore emphasized site-specific habitat protection for individual animals rather than managing for viable populations. However, maintenance of viable populations more directly addresses ESA's goals.

To enforce take prohibitions, regulatory agencies often attempt to maintain databases that map the occurrences and the annual locations of known individuals. Because the focus has been on determining point locations for preserving an individual's habitat (i.e., to protect against incidental take), these data have not been fully utilized to determine population productivity and further define important habitat relationships. The prevailing record-keeping system has some utility when the subject species returns annually to a prominent breeding area. For example, bald eagle nesting areas are effectively protected with this type of system. For obvious reasons, this system does not work for species that are difficult to detect and have poorly documented habitat associations.

Regulations focusing on individual animals produce powerful disincentives that can actually result in accelerated habitat loss. As an example, incidental take guidelines for northern spotted owls on private lands required that at least 40 percent of the median home range be maintained in suitable habitat around known nests. Median home ranges were considered to be circular with a radius of 1.8 to 2.2 miles (variable by region). Where these owl circles included more than one owner, the landowner first able to secure a state forest practices permit immediately harvested much of the suitable owl habitat down to the 40 percent threshold. Upon establishment of a regulatory site centre, landowners had a powerful incentive to submit immediately forest practices applications and then harvest surplus habitat as fast as possible. After all, if one landowner hesitated, then the opportunity to realize income from timber harvesting within that circle was used by an adjacent landowner.

When an individual occurrence is known and incidental take guidelines require habitat preservation at the known site, there are few incentives for landowners to maintain suitable habitats not yet known to be occupied. Other than owl circles, for example, the incentive has been for landowners to harvest stands of habitat that appear to be suitable but not yet occupied. At best, these types of regulations result in isolated preserves of suitable habitat that will remain only as long as known individuals occupy the sites. The same perverse incentive has been blamed for systematic elimination of critical habitats of listed species on private forest lands in several regions of the United States (Stroup 1995; Wilcove et al. 1996).

Implementing ESA on private lands often results in disincentives for developing or maintaining habitats for unlisted species. Although unlisted species are not afforded direct conservation benefits under ESA, some may have suffered from the same disincentives that have adversely affected habitats of listed species. Fear of future regulations may cause some landowners to avoid reporting known occurrences of species yet to be listed. The same uncertainty may lead to the premature harvesting of important forest habitats. We may be forgoing future management flexibility for some taxonomic groups on private lands by not providing landowners with a process for addressing conservation concerns until it is too late (i.e., before the species becomes listed).

Vertebrate groups such as forest-floor small mammals, stream-breeding amphibians, and primary cavity excavators might not have the charisma associated with spotted owls and pacific salmon, but they are nevertheless important components of biological diversity in the forests of the US Pacific Northwest. With some planning efforts, the habitat conservation needs of these species can be effectively and efficiently accommodated on managed private forest lands. However, the planning processes must be in place to ensure that adequate incentives are provided. These incentives must capitalize on a landowner's self-interest.

The USFWS has recently recognized landowners' fear that habitat enhancement efforts may eventually constrain use of their lands and acknowledged that this disincentive is an obstruction to recovery. With the urging of the Environmental Defense Fund and others, the USFWS has sought to secure agreements with some landowners by offering "safe harbour" in exchange for affirmative action to preserve or enhance habitats beyond what is required to avoid incidental take. Assurance of safe harbour forgoes federal constraints on future use of their lands should they ultimately be inhabited by a listed species (Wilcove et al. 1996). Similar assurances for species unlisted at present could result in better conservation efforts on private lands.

The Promise of "Ecosystem Management"

The problems defined here are largely the pathologies of traditional command-and-control regulatory processes. For private lands, the awkward processes created to avoid incidental take hinder progress in meeting ESA goals. These problems do not necessarily imply that ESA has been a failure. In fact, the scientific principles of the act are judged by many scientists to be well grounded (Ecological Society of America 1995; National Academy of Sciences 1995).

A process that focuses on managing forest habitats across multiple spatial scales over relatively long planning periods could be more effective for meeting ESA goals. Some observers have suggested that this is best attained through a coarse-filter, ecosystem-based approach (Noss, LaRoe, and Scott 1996; Wilcove et al. 1996). This system assumes that needs for most species and ecological processes are met by conservation of land areas and representative habitat types. The coarse-filter approach is most appropriate for conserving species that are not yet rare and/or about which we know little. For species that fall through the gaps of a coarse-filter approach, a more traditional fine-filter approach is appropriate. This latter approach includes the intensive species-specific management and monitoring that have traditionally been the operational role of management under ESA.

Proposals advocated by various policy makers have included the coarse-filter approach in a process known as ecosystem management. Federal policy makers are now pursuing a fundamental shift to ecosystem management as a means to implement ESA (Miller 1996; Morrissey 1996; Patlis 1996). But debates on the scientific underpinnings and social consequences of such a fundamental shift have slowed the implementation of ecosystem management.

As refinements of the scientific underpinnings of managing ecosystem functions have progressed among ecologists (see Christensen et al. 1996; Kaufmann et al. 1994), some managers have struggled with applications of the various processes in real landscapes. As applications are constructed, a major shift in thinking is required to balance the uneasy tension between two different scientific approaches to explaining variability within natural communities. The first approach is a reductionist science of parts; the second is the science of integrating those parts. The reductionist approach uses experiments and traditional hypothesis testing to provide resolution of processes; the integrative approach uses models and adaptive decision making to manage the uncertainty of complex systems (see Holling 1996). Because uncertainty cannot be eliminated when managing ecological systems, it is precisely this promise of being able to manage uncertainty that initially attracted ecologists and managers to ecosystem management.

But there is more to ecosystem management than ecology and landowner economics. Some have argued that the social sciences are actually more important than ecological disciplines in making ecosystem management work (Roe 1996). Those who argue for this approach visualize ecosystem management more as a participatory process among various stakeholders, with local leaders and residents being the experts who initiate and guide the planning process (Roe 1996).

How private landowners might be able to participate in ecosystem management is not yet fully understood. Ecosystem management's increased emphasis on participatory decision making and integrative management of regional landscapes (e.g., Noss 1983) causes apprehension that the process may be a new means to transfer private property to a common pool (Gidari 1995). On private forest lands, there may be additional legal concerns. For example, planning efforts encompassing multiple ownerships could require competing landowners to share information about anticipated harvest levels and to agree on the timing and location of a future timber harvest. In addition to a potential loss of advantage by revealing confidential information, this would present liability concerns under antitrust laws (Pauw, Green, and Ross 1993).

Depending on how the process is developed, ecosystem management could merely replace the specific unintended consequences of prevailing processes with new and different disincentives. To avoid this trap, these new planning processes must not be directly at odds with private property ownership, they must not serve to expand rigid bureaucratic processes, and they should translate conservation value into economic value (Olson 1996). This economic value does not necessarily translate into direct payment incentives to private landowners, although it could. The challenge is to define planning processes that (1) remove the current disincentives associated with enforcing ESA take avoidance; (2) use a suitable combination of coarse-filter and fine-filter management approaches; (3) use site-specific knowledge to determine appropriate conservation measures; (4) delegate compliance and effectiveness monitoring to on-site land managers – implementation being subject to verification by regulatory agencies; and (5) capitalize on self-interest by providing regulatory certainty for the duration of a planning period.

Habitat Conservation Planning

Assurance of a future ability to harvest timber is one of the strongest incentives that a private forest landowner can receive. If the regulatory future is uncertain, then the landowner's economic future is less certain. To gain some regulatory certainty, several forest landowners in the US

Pacific Northwest have implemented management plans designed to conserve habitats of federally listed species. These plans were negotiated with the USFWS and NMFS under the authority of subsection 10(a) of ESA. The habitat conservation planning (HCP) process was authorized in a 1982 amendment to ESA to allow incidental take of listed species associated with nonfederal activities. HCPs are required to support issuance of an incidental take permit allowing a landowner's otherwise lawful activities to continue as long as all resulting incidental takings of listed species are minimized or mitigated to the maximum extent practicable.

Only fourteen incidental take permits were completed during the period 1983-92. The rate of HCP preparation accelerated during the succeeding four years, so that by August 1996 a total of 179 permits had been issued. Until recently, HCPs were largely negotiated for single species, many of which were associated with land-use conversion and construction projects.

Only since 1992 has the HCP program been focused on private forest lands. During this time, the northern spotted owl has been the primary subject of four single-species HCPs in Washington, Oregon, and northern California (Murray Pacific Corporation 1993; Oregon Department of Forestry 1996; Simpson Timber Company 1992; Weyerhaeuser Company 1995). Because forest habitats require several decades for management and development, these HCPs were written for relatively long periods, typically from fifty to 100 years. Even though these plans provide relief from incidental take prohibitions for northern spotted owls, they do not provide protection or conservation measures for subsequently listed species. For this reason, the most recent HCP efforts have sought to provide conservation measures for all species of potential concern during the life of the plan. This multispecies treatment requires extensive assessments, conservative planning efforts, and intensive monitoring commitments. In exchange for management commitments, private landowners entering into multispecies HCPs are given assurances that their actions will not be further restricted (US Fish and Wildlife Service and National Marine Fisheries Service 1996).

The idea that a deal is a deal was embodied in Secretary Babbitt's 1994 "no surprises" policy. Under no surprises, landowners operating in good faith under a properly functioning HCP will not be required to provide additional mitigation for species covered under the plan. If additional mitigation is required on their lands, it will not be at their expense – except for certain extreme circumstances. No surprises has received criticism from some environmental groups, which claim that the policy negates adaptive management. If artfully implemented, however, adaptive management and no surprises may not be mutually exclusive (see the following case study). It is this no surprises policy that has provided the incentive for private forest landowners to undertake multispecies planning efforts. Many private forest landowners would not have come to the negotiating table without this demonstration of good faith by USFWS and NMFS.

Port Blakely's Habitat Conservation Plan

On 19 July 1996, Port Blakely Tree Farms finalized a fifty-year HCP for a 7,486 acre parcel of land under its ownership in Pacific and Grays Harbor Counties, Washington. Forests on the plan area are dominated by stands of second-growth Douglas-fir and western hemlock fifty to sixty-five years old. The area contained suitable habitat within the median home range of two northern spotted owl site centres. There are over forty-five miles of perennial streams traversing the area, more than half of which serve as spawning and rearing habitats for Pacific salmon and resident trout. I was a member of the company's HCP team that devised a conservation strategy, incorporating harvest scheduling, silvicultural treatments, riparian management buffers, and prescriptions for avoiding mass waste and control of surface erosion.

We used an intensive forest habitat and stand inventory system within a geographic information system (GIS) accompanied by a simple growth model to project habitat conditions across the plan area throughout the planning period. We then set target habitat conditions for terrestrial and riparian habitats and determined appropriate time intervals to reach such conditions. Specific conservation measures such as thinning and retention of special habitat features in commercial second-growth forests are used to accelerate development of habitat characteristics commonly associated with late-successional or old-growth forests. Other management prescriptions are used to protect and maintain the function of aquatic and riparian habitats important for conserving several fish and amphibian species. These overall management regimes are coarse filters for conserving fish and wildlife species commonly associated with late-successional forests, streams, and riparian habitats.

Given the development of various habitat types, the team was able to optimize overall habitat conditions with Port Blakely's timber harvest goals during the fifty-year planning period. By further focusing on habitat features known to be critical for important prey species such as northern flying squirrels (*Glaucomys volans*), we were able to refine a timber management and harvest plan that mitigated for any short-term incidental take of northern spotted owls that could occur as a result of commercial timber harvesting. Under the final strategy, we projected suitable foraging habitat for northern spotted owls to remain at or above current levels for the first fifteen years, followed by a decline to the year 2025. At that low point, stands more than eighty years old will account for about 77 percent of the total suitable habitat. Foraging habitat then increases to about 70 percent of the planning area by the end of the planning period (Figure 8.1).

Port Blakely's HCP supports an incidental take permit for northern spotted owls, marbled murrelets, bald eagles, and peregrine falcons. More importantly, the HCP's implementing agreement provides assurances that, for the full life of the plan, additional mitigation will not be required for any future listings of species associated with habitats on the plan area. Conservation measures address long-term habitat management for forty-two of those species in a manner that mitigates for potential incidental take at a rate that the environmental assessment determined would exceed the rate of potential incidental take. The forty-two species include all federal and state candidates for ESA listing as well as important indicator species of additional conservation concern. This provision constitutes a form of mitigation banking that not only provides a conservation benefit for those species of concern but also provides Port Blakely's ownership with regulatory certainty with regard to ESA. The disincentives provided by the previous command-and-control process have been removed, and Port Blakely's planned timber harvests and associated activities may proceed as long as the conservation measures of the HCP remain as part of routine forest management.

Figure 8.1

Proportion of Port Blakely's Habitat Conservation Plan area as suitable habitat for northern spotted owls

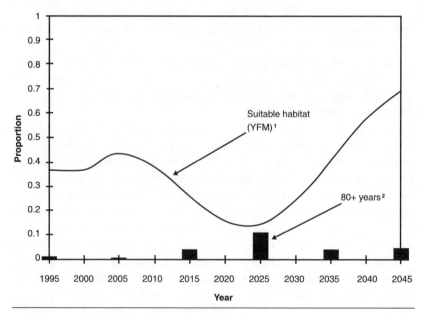

1 YFM = young forest marginal
2 80+ years = forest in 80+ year age class during the fifty-year planning period (1995-2045)

Managing Uncertainty

By character, wildlife species and their ecosystems are often inconspicuous, making them hard to locate and monitor. This complicates evaluation of any conservation program, making it difficult to claim success or failure of a given effort. Because managers rarely have infallible knowledge of species and ecosystem responses to specific management actions, they remain ignorant and hope for the best, or they learn to achieve objectives by adaptive management or monitored trial and error (Walters and Holling 1990). From a policy perspective, an HCP is more defensible (i.e., it has a higher likelihood of long-term success in meeting conservation objectives) if the land manager provides an overriding model for dealing with the uncertainties associated with long-term planning. This is the process for using ecological and operational information to make the changes needed to effect the overall conservation objectives.

In most cases, factors controlling occurrence and abundance of wide-ranging species are not manageable on single ownerships. In practice, therefore, it makes more sense for small to mid-sized landowners to concentrate their efforts on managing and monitoring features that contribute to suitable habitats for various species. This is in contrast to the potential for extremely large landowners to manage and monitor populations – perhaps with maintenance of certain population levels as an effectiveness target. The implementation process described here may be most operable at small (100-1,000 ha) to mid-sized (1,000-10,000 ha) spatial scales, in which the objective is maintenance of viable populations over relatively short time periods (up to fifty years). With modifications, the concept could work for larger land holdings.

The process for managing uncertainty under the Port Blakely HCP is illustrated in Figure 8.2. This process requires that a landowner invest in appropriate levels of compliance and effectiveness monitoring. *Compliance monitoring* is simply an accounting and reporting effort that verifies that management prescriptions have been implemented. *Effectiveness monitoring* is the gauge by which the outcome of implementing the management prescriptions is assessed. Effectiveness monitoring efforts are generally designed with some statistical control so that management effects can be distinguished from other changes that occur over time (e.g., succession, natural disturbance). The appropriate level of effectiveness monitoring depends on the relative level of uncertainty that the *management prescriptions* (i.e., conservation measures) will result in *target conditions*.

Research or *validation monitoring* should be designed with an experimental approach to determine species response to various habitat conditions and/or management actions. The work does not necessarily need to be conducted by the individual landowner. However, monitoring should ensure that results used for setting target conditions are applicable to the

Figure 8.2

Habitat conservation monitoring process

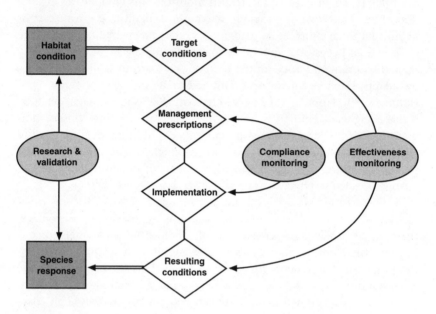

area to be managed and that future research will be conducted for those species-habitat relationships that are important for refining currently un-reliable models. Appropriate levels of research or validation monitoring will depend on the reliability of current knowledge of the relationship between *species response* and *habitat condition*.

The link between species response and habitat condition is inferred from research results. Target conditions to provide those habitat condi-tions are then defined, and management prescriptions (i.e., conservation measures) are agreed upon. Implementation is followed by compliance monitoring to verify fulfilment of prescriptions. The resulting conditions are later compared to target conditions through effectiveness monitoring. To complete the sequence, further research and validation monitoring are used to associate *species response* to the resulting conditions. If necessary, incremental adjustments in target conditions, management prescriptions, and implementation techniques are made throughout a planning period.

This monitoring process supplies opportunities for incremental corrections to increase the likelihood of a successful management plan – thus, it adds rigour to the overall management strategy. For example, it may be determined through effectiveness monitoring that the resulting conditions for a set of habitat features have not met (or are not likely to meet) the target condition by the time projected in a plan. If so, then the manager would determine whether the deficiency is due to short-comings of either implementation or management prescription. If com-pliance monitoring verifies that prescriptions were implemented correctly, then management prescriptions should be changed to increase the certainty that the target condition is met. On the other hand, if compli-ance monitoring indicates shortcomings in implementation, and the management prescription remains reasonable, then stronger compliance accountability would be required. As more reliable knowledge is gained through research and validation monitoring, target habitat conditions may need to be adjusted. As a policy matter, the original management plan should state the level of responsibility for meeting certain habitat needs so that new knowledge can be appropriately used in adjusting target conditions.

Implementing several different management prescriptions, all of which could result in the target conditions, could strengthen the process. Where there are collections of similar units (e.g., forest stands, riparian areas, dis-tinct animal occurrences) that could be managed independently, each alternative prescription could then be applied in multiple units or experi-mental replicates. Use of experimental replicates would provide opportu-nities for active adaptive management (Walters and Holling 1990).

Obviously, this approach does not account for large-scale disturbances that would influence population viability at the regional level (i.e., metapopulations). Planning for unanticipated catastrophes is an entirely different type of risk and uncertainty assessment.

A Leap of Faith for Landowners

Port Blakely's HCP applies only to a small portion of the company's overall forest land holdings and was undertaken as a leap of faith by the ownership. Consummation of the agreement bound the company to a monitoring and management feedback process. Briefly, this process com-mits the landowner to monitor its compliance with all conservation meas-ures of the HCP and to report to federal agencies on a biennial schedule. Effectiveness of conservation measures in meeting stated goals is continu-ally monitored and reported to the agencies every five years. Port Blakely's performance under the HCP is further reviewed by the agencies through verification monitoring. If effectiveness monitoring does not indicate habitat trends as expected, then Port Blakely is committed to adjust its implementation of conservation measures to ensure projected habitat conditions are achieved.

Two other multispecies HCPs are being implemented on private forest lands in Washington (Murray Pacific Corporation 1993; Miller 1998), and several others are in negotiations with USFWS and NMFS. The willingness of these landowners to enter into these agreements is directly attributable to the visionary actions of several administrators within the two federal agencies that administer ESA.

The HCP processes require input from forest economists, silviculturists, logging engineers, loggers, and other forest workers who are not typically exposed to the reasoning behind habitat management recommendations proposed by fish and wildlife biologists. Not only do these individuals gain a more thorough understanding of the habitat needs of various species, but also they are more likely than before to incorporate these needs into other management projects. Equally important is the education that fish and wildlife biologists receive from these other professionals, thus increasing their ability to devise management options that not only contribute to conservation but are also more likely to be effectively integrated into the planning efforts of other forest professionals.

Although these agreements signify substantial progress in overcoming the disincentives of command and control authority of ESA on private lands, the HCP process as it currently exists may not be suitable in all situations. The current process is more likely to be successful when tenure of ownership is stable, where there are established and predictable markets for forest products, and where ownership sizes are large enough to include a range of developing habitat conditions over time. The effectiveness of HCPs in conserving fish and wildlife resources will also depend on the willingness of landowners to validate and modify existing plans given the results of monitoring and research.

A Cultural Change for Agencies

The procedures and policies that govern the negotiation of HCPs and similar agreements are still being formed. State agencies, environmental groups, and Native American interests are not yet entirely satisfied with the level of participation that they have been afforded by the HCP process. Private landowners must now work to ensure that these groups are given genuine opportunities to participate.

Agencies responsible for protecting fish and wildlife resources have traditionally played a reactionary role when dealing with conservation issues on private lands. If private land conservation efforts are to be successfully implemented, then agencies must not take a wait-and-see attitude toward ownership-specific conservation plans. To make further progress, state agencies and other organizations must undergo a cultural change and restructure their thinking. They must approach problems aggressively and with the priority of protecting biological diversity leading their actions – not the sustenance of existing bureaucratic processes. This may mean making trade-offs and taking actions considered suboptimal in the short term (i.e., incidental take of listed species) in order to improve future conservation (Mangel et al. 1996).

Acknowledgments

I would like to thank Debra Salazar and Don Alper for their comments and encouragement, as well as Lisa Moss and Jim Paul for their thoughtful comments and suggestions on an earlier draft of this chapter. Thanks also go to Court Stanley and Jerry Bailey for their continued perseverance in managing Port Blakely's HCP with an eye toward wildlife conservation.

Note

1 See *Bruce Babbitt, Secretary of the Interior, et al.* v. *Sweet Home Chapter of Communities for a Greater Oregon et al.,* 115 S.Ct. 2407 [1995].

References

Anderson, T.L., and D.R. Leal. 1991. *Free Market Environmentalism.* Boulder, CO: Westview Press.

Christensen, N.L. et al. 1996. "The Report of the Ecological Society of America Committee on the Scientific Basis for Ecosystem Management." *Ecological Applications* 6(3): 665-91.

Ecological Society of America. 1995. *Strengthening the Use of Science in Achieving the Goals of the Endangered Species Act.* Washington, DC: Ecological Society of America.

Gidari, A. 1995. "The Economy of Nature, Private Property, and the Endangered Species Act." *Fordham Environmental Law Journal* 6(3): 661-87.

Hilborn, R. 1996. "Do Principles for Conservation Help Managers?" *Ecological Applications* 6(2): 364-65.

Holling, C.S. 1996. "Surprise for Science, Resilience for Ecosystems, and Incentives for People." *Ecological Applications* 6(3): 733-35.

Kaufmann, M.R., et al. 1994. "An Ecological Basis for Ecosystem Management." USDA Forest Service, General Technical Report RM-246.

Mangel, M. et al. 1996. "Principles for the Conservation of Wild Living Resources." *Ecological Applications* 6(2): 338-62.

Miller, G. 1996. "Ecosystem Management: Improving the Endangered Species Act: Perspectives on Ecosystem Management." *Ecological Applications* 6(3): 715-17.

Miller, M. 1998. "Plum Creek Habitat Conservation Plan." *Improving Integrated Natural Resource Planning: Habitat Conservation Plans* [web page]. National Center for Environmental Decision-making Research. www.ncedr.org/casestudies/hcp/plum.htm [14 October 1998].

Morrissey, W.A. 1996. "Science Policy and Federal Ecosystem-Based Management." *Ecological Applications* 6(3): 717-20.

Murray Pacific Corporation. 1993. "Habitat Conservation Plan for the Northern Spotted Owl on Timberlands Owned by Murray Pacific Corporation in Lewis County, WA." Murray Pacific Corporation, Tacoma, WA.

National Academy of Sciences. 1995. *Science and the Endangered Species Act.* Washington, DC: NAS Press.

Noss, R.F. 1983. "A Regional Landscape Approach to Maintain Diversity." *BioScience* 33: 700-06.

Noss, R.F., E.T. LaRoe, and J.M. Scott. 1996. "Endangered Ecosystems of the United States: A Preliminary Assessment of Loss and Degradation." Unpublished report, National Biological Service.

Olson, T.G. 1996. "Biodiversity and Private Property: Conflict or Opportunity?" In W.J. Snape III (ed.), *Biodiversity and the Law.* 67-69. Washington, DC: Island Press.

Oregon Department of Forestry. 1996. "Elliot State Forest Habitat Conservation Plan. Coos District, OR." Oregon Department of Forestry, Salem.

Patlis, J. 1996. "Biodiversity, Ecosystems, and Endangered Species." In W.J. Snape III (ed.), *Biodiversity and the Law.* 43-58. Washington, DC: Island Press.

Pauw, J., T.J. Green, and D.C. Ross. 1993. "Balancing Endangered Species Regulation and Antitrust Law Concerns." Washington Legal Foundation, Working Paper Series No. 54.

Roe, E. 1996. "Why Ecosystem Management Can't Work without Social Science: An Example from the California Northern Spotted Owl Controversy." *Environmental Management* 20(5): 667-74.

Rolston III, H. 1991. "Life in Jeopardy on Private Property." In K.A. Kohm (ed.), *Balancing on the Brink of Extinction: The Endangered Species Act and Lessons for the Future.* 43-61. Washington, DC: Island Press.

Simpson Timber Company. 1992. "Habitat Conservation Plan for the Northern Spotted Owl on the California Timberlands of Simpson Timber Company." Simpson Timber Company, Arcata, CA.

Stroup, R.L. 1995. "The Endangered Species Act: Making Innocent Species the Enemy." *PERC Policy Series* 3: 1-19.

US Fish and Wildlife Service and National Marine Fisheries Service. 1996. "Endangered Species: Habitat Conservation Planning Handbook." Unpublished report.

Walters, C.J., and C.S. Holling. 1990. "Large-Scale Management Experiments and Learning by Doing." *Ecology* 71(6): 2060-68.

Weyerhaeuser Company. 1995. "Habitat Conservation Plan for the Northern Spotted Owl. Millicoma Tree Farm. Coos and Douglas Counties, OR." Weyerhaeuser Company, Federal Way, WA.

Wilcove, D.S., M.J. Bean, R. Bonnie, and M. McMillan. 1996. "Rebuilding the Ark: Toward a More Effective Endangered Species Act for Private Land." Environmental Defense Fund, unpublished report.

9
Multistakeholder Processes: Activist Containment versus Grassroots Mobilization[1]
Mae Burrows

During the 1980s and 1990s, governments discovered multistakeholder processes as an effective tool for dealing with difficult issues such as land-use conflicts. In 1991, at the height of "the war in the woods" between environmentalists and logging interests in British Columbia, the provincial government organized a multistakeholder process, the Commission on Resources and the Environment (CORE), to bring various interests together in an attempt to resolve land-use conflicts related to logging.

This chapter begins by briefly explaining the social context of the CORE process as well as the structure of, and participants in, the process. I then isolate the elements of an effective multistakeholder process. I evaluate CORE by applying these elements to my participant-observer research conducted over a five-year period. The chapter concludes with a series of recommendations for activists to consider before deciding to participate in alternative dispute resolution or multistakeholder processes.

The Social and Political Context of CORE
Historically, the BC economy has been dependent on the exploitation of natural resources, especially logging, mining, and fishing. Ninety-four percent of the forest land base in the province is publicly owned Crown land. Since the mid-1940s, Crown land in British Columbia has been leased to forest companies that are allowed to cut at rates set by the provincial forester (see the discussion of tenure arrangements in the province in Chapter 2). By the 1970s, public awareness of the amount and type of logging taking place had grown enormously. Damage to the environment attributed to logging included loss of old-growth ecosystems, destruction of fish habitat from landslides and siltation, and overcutting of timber. These problems were officially recognized by the provincial government's Royal Commission on Forest Resources in 1976 (Pearse 1976). Furthermore, many believed that government management of forest

practices was ineffective in protecting resource values such as fisheries and recreation. Critics maintained that contemporary practices were effective only in supporting the economic goals of large, integrated forest companies (Pinkerton 1993: 34).

Increasing awareness of forest practices and their consequences prompted new forms of political activism. They included mass protests, tree climbing to prevent logging, and blockades of public and logging roads. By the late 1980s, conflict and acrimony around forest practices had escalated to such an extent that many were calling it "the war in the woods," said to be fought "valley by valley" and pitting preservationists against resource extractors.

A more aware public demanded a voice in decisions about how much and what kind of logging took place in the remaining pristine watersheds and in the establishment of long-term land-use planning. Government officials, alarmed by the deepening polarization, sought different ways to deal with the discord and looked to nonhierarchical ways of making land-use decisions. Consensus processes were seen as alternative forms of dispute resolution. They were being used in many forums, including disputes involving uranium mining in Colorado, conflicts between highway interests and transit system advocates in New York City, logging battles in Montana, and disputes among timber, fish, and wildlife interests in the State of Washington. In 1991, the New Democratic Party (NDP) won the provincial election in British Columbia, and one of its first items of business was to address the high level of conflict between two key constituencies – environmentalists and loggers. In 1992, the NDP government established the Commission on Resources and the Environment (CORE), with a mandate to develop a provincial land-use strategy.

Research Methodology and Involvement

When I began this research, I was director of the T. "Buck" Suzuki Environmental Foundation, an organization that undertakes environmental work related to fish habitat and water quality. The foundation works closely with the commercial fishing community, particularly with members of the United Fishermen and Allied Workers Union (UFAWU). As part of my assigned work, I participated as one of the fishery representatives on the Vancouver Island CORE, one of three BC "hot spots" that became the subject of a CORE process. I reported to the members of the Commercial Fishing Industry Council (CFIC), a federally sponsored umbrella organization representing vessel owners, processing companies, organizations involved with various gear types, and UFAWU, which represents both fishers and shore workers. Much of the information contained in this chapter is based on my participant observation, both indirectly through field interviews during the year in which the process occurred and directly through formal interviews with representatives from each sector following conclusion of the process. The indirect observations include the above-mentioned contacts as well as discussions with people from NGOs and fishing groups and individuals who had chosen not to participate in the process. I received all minutes of CORE meetings, newsletters and other written material on CORE, and an extensive file of newspaper clippings.

This study analyzes the extent to which in multistakeholder processes the stated intention of "empowerment" – granting equal authority and power to participants who in reality are not equal in terms of financial resources, education, public influence, formal position, social class, control of land, and control of production – can be realized. The chapter examines critically whether there can be "shared decision making" given power imbalances among participants in the process.

On one side were forest companies with legally entrenched rights to the land. On the other side were various interest groups with weak or nonexistent power bases from which to affect the negotiations. I wanted to examine this model of consensus negotiation, which included both powerful and powerless groups seen by the government to have legitimate interests or stakes in the issue. Could multistakeholder consensus negotiations really bring all interests together and encourage them to engage in rigorous, substantive debate about their differences? Or would the fact that the government had initiated and structured CORE serve to contain debate and keep the conflict out of the media? Using the interest-based stakeholder model also meant that the public (nonaffiliated individuals not formally associated with interest groups) was excluded from the discussions. I wanted to analyze what this exclusion meant in terms of the claim that the CORE model provided a more inclusive forum than did traditional forms of dispute resolution.

I also wanted to examine the intentions of participants and their stated and unstated goals. One question was whether some individuals engaged in CORE with the intention of gaining greater understanding and finding common ground, while others participated only to achieve strategically predetermined goals. How would consensus negotiations be affected when participants with different intentions took part?

Participants in the CORE process expressed different value systems and world outlooks, and I wanted to discover if consensus was even possible among people with such different perspectives. Could a consensus-based roundtable process contribute to a shift in values as a result of contact between groups that wouldn't normally have contact with one another?

I also had a pragmatic reason for undertaking this study. Simply put, should environmental organizations participate in roundtable consensus forums at all, or should they concentrate their efforts in other arenas, such as grassroots organizing, direct political lobbying, public education, and media campaigns? Which approach would be more beneficial to groups like mine, the T. "Buck" Suzuki Environmental Foundation?

The Structure and Content of CORE

In 1992, Premier Mike Harcourt announced that he was setting up CORE as an alternative dispute resolution process. It was touted as a new way to make land-use decisions. The government appointed the provincial ombudsman, Stephen Owen, as commissioner, and he then assembled a staff predominantly of lawyers and planners. This staff determined that CORE would be a mediated, multisectoral, interest-based consensus negotiation.

Participation in CORE would occur through "sectors," meaning that each seat at the table would be allocated to a major stakeholder or government group with an interest in the outcome of land-allocation decisions. The sectors would not necessarily be pre-existing groups; in some cases, they would consist of organizations that had never worked together. Nor did they represent purely geographical areas. The prime criteria for selecting sectors would be "effectiveness" and "inclusivity" (CORE 1992: 10). The view was that too many groups or individuals sitting at the table would make it ineffective, yet all major interests had to be represented. Staff produced guidelines stating that "participant groups should structure themselves so as to have a broadly representative voice, effectively including every element of their constituency" (CORE 1993: 3). The process by which sectors gained a seat at the table was neither clear nor consistent. Some groups were invited to form a sector, while other sectors with less clearly defined constituents took longer to organize themselves and gain a seat at the table. The sectors that ended up at the CORE table were agriculture, conservation, forest employment, forest independents, forest managers and manufacturers, fishery, general employment, local government, mining, outdoor recreation, provincial government, social and economic sustainability, tourism, and youth. That the provincial government was to constitute a sector was predetermined by CORE staff. This decision caused many problems.

The government representative was a prominent NDP lawyer on contract to cabinet to attend CORE meetings and report to cabinet. He was the only government spokesperson at the table, although he would consult with technical staff from the Ministries of Environment, Forestry, and Municipal Affairs. Provincial government was a problematic sector insofar as the representative spoke for the government's interests at the table at the same time that he was reporting to cabinet, which was responsible for making decisions based on the outcome of the negotiations. The government sector's anomalous position is reflected in its interest statement: "Government has a unique three-faceted role at the Table in that it participates as a Sector with interests in the negotiation, is responsible for review and approval of the Table's recommendations, and must develop, fund and implement the final determination (plan)" ("Government Interest Statement," CORE 1994).

The government sector representative's job included "improving public understanding of the objectives and impacts of new resource initiatives, such as the Protected Areas Strategy" ("Government Interest Statement," CORE 1994), as well as being a conduit to cabinet for recommendations coming from the table, such as the strategy for economic transition. Yet the representative was also an active negotiator at the table.

Participants were informed that interest-based representation and agreement by consensus would be intrinsic to the CORE process. CORE materials defined what was intended as, "literally, 'feeling together,' the process of coming to a communally acceptable agreement through group participation in formulating the outcome" (CORE 1992: 40).

A number of sensitive issues were excluded from the agenda. They included clear-cutting practices and the land tenures on Crown land of multinational forest companies. Also, some geographical areas within the regional boundaries were arbitrarily excluded: north and south gulf islands, south and midcoast forest districts, and Clayoquot Sound.

After nearly a year of two-to-three-day-long monthly meetings, the government-imposed deadline of 23 November 1993 arrived without consensus on many major issues. It was left to the CORE commissioner to prepare his own recommendations to cabinet. Following the release of his report, the logging community responded negatively by holding mass-protest rallies characterized by yellow ribbons and "No to CORE" signs. Despite hundreds of hours invested by volunteers, CORE did not produce an agreed-upon land-use strategy that various communities could live with. Many believed that the significant decisions were made elsewhere.

The Alternative Dispute Resolution Literature

What went wrong with the Vancouver Island CORE? To answer this question, examining CORE in the context of alternative dispute resolution (ADR) scholarship is critical. The scholarly literature on ADR reveals conditions that make for effective multistakeholder ADR processes. From this literature review, we can gather criteria with which to measure the effectiveness of CORE.

While ADR is not a panacea for resolving difficult land-use issues, it has worked in many situations. Lawrence Susskind enumerates four principles – "fair, efficient, wise and stable" (cited in Bowering 1992: 5) – that should be applied in judging the success of ADR. First, at the conclusion of a successful process, all parties should feel that they were treated fairly (Susskind, cited in Bowering 1992: 5; Cormick 1987: 41). Second, the process should be efficient in terms of both time and money and ensure that potential advantages to all participants were not left unclaimed. Third, the agreement reached should be stable. And fourth, according to Susskind, there should be a "wise design, a wise development package," which means that, when we look back on the process in the future, we will consider it to have been a good one (cited in Bowering 1992: 5).

There are situations in which ADR is difficult to apply. An important caveat to the use of ADR in British Columbia is the reality that most disputes are played out on Crown land. As Gunton and Flynn (1992: 15) point out, "The application of ADR to crown land is particularly challenging. Crown land planning often involves many parties, the issues are poorly defined, there is often an incentive for some parties to avoid a decision in order to maintain the status quo and there are often fundamental value differences."

To be successful, ADR must meet a number of criteria. Five themes predominate in the literature.

1. The Process Must Be Designed by the Participants

A point frequently made is that the process must be designed by the participants. For both the British Columbia Round Table on the Environment and the Economy and the National Round Table, the point has been made that "it is critical that all parties have an equal opportunity to participate in designing the process" (National Round Table on the Environment and the Economy [NRTEE] 1993: 8). This process should include developing a "structure for how the process will work, including meeting formats, working sub-groups, caucuses, and ground rules" (NRTEE 1993: 77, 14). Susskind agrees that the process should be "ad hoc": "there should be room in each situation for the participants to design the dispute resolution process they prefer" (Susskind and Cruikshank 1987: 77). And Cormick adds that, "generally, if those who are expected to use a dispute settlement process do not see the need for such a mechanism or are not involved in its design and implementation, it is unlikely to work ... Evidence suggests that the least promising way to develop a system is to import it from elsewhere" (1987: 41).

2. The Agenda Must Be Determined by the Participants

Participants must also develop the agenda in order to feel that they have been treated fairly and to improve the chances of achieving a stable agreement at the conclusion of the process. If an agenda has been predetermined, Amy cautions, the process may be used to legitimize and implement policies and decisions that have been made elsewhere. He points to a US government report that encouraged federal officials to use "conflict management tools like conciliation, facilitation and mediation ... to avoid regulatory stand-offs and move opponents to mutually acceptable settlements" (1987: 151). This report raises the possibility that an ADR can be used as a management tool rather than as a participant-run process: conflict management "should not be confused with the requirement for public participation. The key word is management. In cases where public groups are fighting with the Federal government, better conflict management means better control over the participation process ... At the outset, the mediator requires the agency to outline its specific constraints – regulatory, political, economic – under which it must operate and within which any final agreement must fall" (Amy 1987: 151-52).

This statement raises the question of differences in definition between facilitation, mediation, negotiation, and management. Facilitation involves assisting parties in the process of achieving a goal. Although mediation is often thought of as an alternative to traditional adversarial approaches, Amy reminds us that this is misleading. Rather, mediation is best understood as an extension of adversarial politics, not a substitute for it (1987: 68). Mediation involves bringing conflicts to the surface and striving for a power balance among disputants. Negotiations are held as part of the mediation process. Management of conflicts, on the other hand, involves control so as to keep certain agendas and points of view submissive. It is important that participants have a keen sense of the kind of process in which they are involved. Participants brought different experiences and definitions to the CORE process, and this diversity caused problems since different definitions reflected different intentions.

By setting the parameters of what can be discussed, the process managers determine what is negotiable. "In essence, the government controls the agenda of the negotiations and predetermines the outcome. In practice this means that although an agency may be willing to grant concessions on details, the basic policy decisions remain nonnegotiable" (Amy 1987: 152). Bachrach and Baratz (1963) call this process "nondecision-making," which they define as "the practice of limiting the scope of actual decision making to 'safe issues'" (632). By limiting the agenda to "safe issues," the process managers ensure that fundamental values and policy issues are not discussed, even if the participants want such issues to be addressed. This narrowing of the issues leads to both an unfair and an unstable outcome.

3. Participation May Be Inclusive or Restricted
Opinion is divided on whether ADR should be inclusive or involve only those with certain kinds of power. Those in the former camp include Cormick, Susskind, and the National Round Table, who say that a guiding principle of consensus processes is that "All parties with a significant interest in the issue should be involved" (cited in Doering 1995: 2). The Environment Canada document *Working in Multistakeholder Processes* goes further to ensure broad public participation: "Invitations can be sent

directly to all known interests asking them to attend and pass on the invitation [to the first meeting] to others. Media such as radio and local television can be used to publicize the meeting. Newspaper advertisements should be run ... posters distributed to public areas such as libraries [and] schools ... or an invitation sent to every household using a flyer delivery service" (Canada 1994: 53).

On the other hand, some students of ADR recommend that only stakeholders who have power or influence to undermine any decision that might be made should be included (British Columbia Round Table on the Environment and the Economy 1991: 22; Flynn 1992: 31). Mediator Harold Bellman argues that participants' access to a dispute resolution process be based on how much power the participants have outside the process: "One of the reasons that mediation works is that it is usually limited to people that have some impact on the situation. I don't ask people who don't have clout to participate in the mediation. This is not public participation, this is cloutful people's participation. That's a real important difference. I don't have anything to do with people without power because they can't affect what I'm doing" (cited in Amy 1987: 134).

If the process managers adopt an inclusive approach to participant involvement, then there is a greater chance that the public – nonaffiliated individuals who have a general interest in the dispute – will have input into the decision. However, given that ADR usually relies on stakeholder interest groups, broader outside interests are excluded from the process. The result may be decisions that are not fair, efficient, or wise. It may be good for the stakeholders but not necessarily for the public interest (Gunton and Flynn 1992: 15). Amy makes the point "that a good mediated agreement almost always satisfies those parties in the negotiations – with little references to the public interest" (1987: 137). Clearly, there is a problem with equating negotiators' actions and interests with the public interest.

Whether the process is formulated according to an inclusive model or a power-broker model, the issue of equality among the participants remains central to an effective and successful program.

4. Funding and Other Resources Must Be Provided to Participants

Inequality is inevitable in the financial resources available to industry and to NGOs. Both the British Columbia Round Table on the Environment and the Economy and the National Round Table of the same name state that participant funding must be provided to stakeholders who need it. As Darling comments with regard to financial resources in the Clayoquot Sound Sustainable Development Task Force, "without such financial assistance, the volunteers (whose involvement was critical to the outcome of the process) were constrained from effective and sustained participation. Funding assistance would also have served to help 'level the playing field' and address power imbalances" (1991: 38). The public at large, as well as certain interest groups, will be disadvantaged in a process such as CORE because of unequal financial resources. Brenneis and M'Gonigle point out that "the general public lacks sufficient resources to participate on an equal basis with affected business interests, necessitating the provision of intervenor funding as a prerequisite to equitable participation in the evaluation of a complex problem" (1992: 9).

There will also be disparities in other resources, including time available, number of individuals involved, friendships and professional relationships, legal rights, access to politicians and bureaucrats, knowledge, negotiating experience and skills, and political power – the influence and power that an individual or group holds outside the process (Cormick 1987; Gunton and Flynn 1992; Wilson 1995; Wondolleck 1988). Susskind considers imbalances in knowledge and skills to be more problematic than imbalances in political power. He suggests that, with regard to the power imbalance, political alliances within the negotiation process can benefit weaker groups, thereby levelling the disparity in power: "Negotiation is not an antidote to inequality. Groups can be, and sometimes are, outnumbered or outmanoeuvred ... On the other hand, we maintain that power in negotiating is dynamic, and that political power away from the bargaining table is not necessarily a good predictor of what will happen once negotiations begin. Coalitions can form, tipping the scales in unexpected ways" (cited in Bowering 1992: 135). On the other hand, writers such as Cohen caution that mediation may be inappropriate "where there is an incorrectable power imbalance" (cited in Lane 1996: 203).

Power balance in a negotiation is also important so that the disputants will participate voluntarily. As Amy points out, "in environmental mediation, the question of voluntariness is a crucial one" and "one of the few institutional safeguards that help to ensure the fairness of the process" (1987: 146, 147). If groups join the process because they believe that they really don't have any other options, then the result will be neither fair nor stable, and the negotiations can become distorted because they have not been undertaken in good faith. As Darling points out, "The process must allow the parties to discuss first with their constituents and then with each other the purpose and desirability of negotiation, their willingness to join the process and the need to involve a mediator. This is fundamental. One of the reasons these processes have failed in the past is that we have forced parties to participate, then we find that they are unable to actually join in a consensus later on" (cited in Roseland 1993: 32).

More important is the incentive that equal power provides for coming to the negotiating table. If, by using their equal power outside the process, the parties have become locked into a stalemate in which each can stymie the other from achieving its goals, then ADR becomes an attractive possibility (British Columbia Round Table on the Environment and the Economy 1991; Flynn 1992; National Round Table on the Environment and the Economy 1993).

However, for participants with unequal power, the pressures to participate are often subtle. When a process is mandated by the government, participants may think that their lack of participation could disadvantage them in future policy decisions or that they will lose public credibility. Amy warns that "outside political pressure can often make it difficult for environmentalists to oppose a mediation effort even when they believe it may be a waste of time" and "that the political pressure to be reasonable can induce environmentalists to stick with a mediation effort even when it is not working" (1987: 177). Gunton and Flynn argue that "successful ADR therefore requires an equitable distribution of power, an assumption rarely validated" (1992: 15).

Why would a group with more power than the others invest its time and energy in the negotiations? Amy suggests that it might engage in mediation even when it has alternatives because through mediation its objectives "would be granted a degree of political legitimacy that would be hard to obtain otherwise" (1987: 149).

5. Differences in Values and Goals Must Be Addressed

Even if ADR meets the criteria discussed in the literature, there is a fundamental issue to address. One tenet of ADR is that there are no right or wrong positions. Auerbach suggests that much of the historical support for consensual dispute resolution has come from tightly knit groups with communitarian values such as Quakers and Puritans. These groups shared values and goals, and "the framework for resolving disagreements was mutual and consensual, not adversarial" (cited in Amy 1987: 83).

However, in the case of environmental conflicts in multistakeholder groups in which communitarian values are not present, ADR can work only when ethics and values are parked at the door. The consensus process frames disputes as being value-free.

Lane states that "it is critically important that the mediator refrain from judging the legitimacy of points raised by the parties" (1996: 201). Portraying environmental disputes, for example, as amoral interest conflicts allows the negotiating forum to accommodate as many interests as possible in reaching a compromise. As Colosi explains, "The mediator's job is to get negotiators to doubt perceptions that block agreement. Those perceptions might be a view of an issue, an understanding of the impact of a proposal, a problem definition, an assumption, or a value ... In this sense environmental mediation is not simply a way of resolving environmental conflicts, it is also a way of redefining the way we think about them" (cited in Amy 1987: 163-64).

While there is the notion in ADR that all parties' views are equally valid, there is also a contradictory notion recognizing that there cannot be a negotiated agreement when fundamental value differences are at stake (Amy 1987; Susskind and Cruikshank 1987; Wondolleck 1998). As a result, Susskind argues, ADR should be restricted to distributional issues (as opposed to fundamental value questions or issues of basic human rights) (Susskind and Cruikshank 1987: 77).

The reality that fundamentally different values and goals exist among participants is often ignored in consensus processes. One CORE participant made this clear: "When you run into a situation where one of the negotiators is a Christian and the other is a lion, the lion says, 'My interest is in eating you,' and the Christian says, 'My interest is in staying alive,' then you have to go a long ways into generalizations before you find common ground" (personal interview).

Bellman makes the important point that "environmental mediation is not an encounter session but an intensely adversarial and combative process, where each side tries to get the most for itself and only compromises when it is forced to" (cited in Amy 1987: 86). Amy also emphasizes "that you can't have consensus among people with different world views" (1987: 184).

From my experience and research, I also conclude that consensus is difficult, if not impossible, to achieve among participants with different value systems. Mansbridge warns how the claim that people have common interests can mislead the less powerful into collaborating with the more powerful in situations beneficial to the latter (Amy 1987: 171). This is why activists engaged in multistakeholder consensus processes must strategically evaluate whether common goals among participants exist. Activists must also evaluate their power base relative to those of others in the process. Even if the consensus process appears to contain the right elements, it may still be significantly unequal.

Although there are no easy solutions to equalizing power among participants in consensus processes, with an awareness of power imbalances participants can better address the problems inherent in such multistakeholder processes.

An Evaluation of CORE

A comparison of CORE with the five themes found in the literature suggests four significant conclusions.

The Agenda, Structure, and Membership Were Predetermined

Because the CORE process had a preset agenda, a predetermined structure, and preferential treatment given to some sectors, many potential participants withdrew, limiting the inclusiveness of the exercise. Friends of

Clayoquot Sound, Greenpeace, and the Western Canada Wilderness Committee chose not to participate for these reasons. Since CORE was a process essentially intended to bring environmentalists and the forest industry together, the fact that these influential environmental groups remained outside was significant. The pre-established conditions restricted full debate.

As Britell points out, "manipulation of citizens' committees begins with the selection of participants" (1992: 21). CORE staff actively invited and assisted the participation of some sectors, while other sectors had to struggle for seats at the table. The "brown" sectors (forest managers and manufacturers, mining, forest employment, and forest independents), and the "green" sector (conservation), were invited to the table at the beginning, and their participation was actively supported by CORE staff (as reported by participants from forest employment, forest independents, and conservation). Other sectors, such as fishery and social and economic sustainability, were given little encouragement to be involved.

Furthermore, that CORE was structured as an interest-based multisectoral forum had been predetermined with no consultation with those who would be involved. When such critical questions are determined outside the process, it is not a self-governed one. Citizens invited to engage in a forum in which the parameters have already been firmly set should be sceptical about the potential for future self-governance in the process. Moreover, any group engaged in a multisectoral forum must closely determine if its participation is considered by the process managers as essential or whether it is included for the disingenuous purpose of making the process look inclusive. This determination will provide strategic information about how to engage in the forum. The group might decide to withdraw if it suspects that the main reason for its involvement is to provide legitimacy to what is essentially a two-party negotiation.

The Playing Field Was Not Level
The assumption inherent in the notion that a roundtable consensus process creates a "level playing field" must be addressed. There are two main components to this myth. The first is that people with vastly disparate power bases, resources, and cultural capital will become equals because, "for a defined period of time on the issues that the participants have agreed to address, they participate as equals" (British Columbia Round Table on the Environment and the Economy 1991: 4). While no process can assume responsibility for creating equality, to ignore inequities in power, resources, cultural capital, and financial support is to ensure that disparities will affect the final outcome of the process. As one participant said, "The problem with the consensus process is that usually it's the vested interests which have the most power and the most ability

to participate, so inevitably their views tend to dominate, and the status quo tends to be perpetuated" (personal interview).

Prior to CORE's beginning, staff decided that multinational forest companies would be included at the table and that the Crown land tenure system and logging practices – by far the two most significant factors in determining land use – would be off-limits for discussion. The extreme power imbalance stemming from the companies' rights to the land undermined any potential for equality in the process. As one conservation sector representative pointed out,

> The forest companies have legally entrenched rights, and that's the tenure system, and the whole idea that every piece of wood in British Columbia has got some company's name on it ... They had more power at the CORE table than other sectors because of the legal rights, and they operate with a completely different mind-set because the fiduciary duty of a board of directors of a corporation is to make a profit for its shareholders. And with that legally bound mandate and fiduciary duty, they have no other way to operate when sitting at a table that tells them "You're going to lose. We're taking from you." So [in the case of] industry, [the] best strategy is to not change. (personal interview)

Rather than being lulled into some degree of misguided confidence that they are being treated equally, people engaged in consensus negotiations need to evaluate critically the strengths and resources of other participants against their own strengths and resources in order to maximize their own effectiveness.

The second component of the "level playing field" myth is that fundamentally opposing views can be accommodated. Even if all parties are somehow treated equally, there is an inherent contradiction in consensus negotiations that people with fundamentally opposed views can agree on substantive issues. Both consensus negotiation and the mediation model assume that the participants have common values and goals as well as equal resources. The lack of commonality in values and resources among the CORE participants was expressed this way: "After going through this process, I don't think people with very different values and interests can actually engage in a consensus negotiation because what the different parties had was so different. The conservationists didn't have the same power base or resources as the forest majors did ... It's an uneven power balance, and you don't have shared values and beliefs. One group values money and their lifestyle, and the other group values the environment" (personal interview).

Another representative made the important point that consensus requires community. Mediation can work in family or neighbourhood disputes – situations in which participants usually share common values and outlooks and in which all the players fundamentally want to solve the conflict to maintain the relationship. Once the hope of common ground and common goals vanishes, the consensus or mediation process will likely dissolve. The consensus process then becomes a strategic negotiation, and many people involved in CORE believed that participants were engaged in strategic rather than communicative action throughout the process. One participant expressed it this way: "It was a while before I realized that negotiating was being done at the table. I think when I realized that it became much easier for me to deal with whatever happened. It became more like a contract negotiation than a consensus procedure" (personal interview). And one of the labour representatives said that "This is about cutting a deal, and anybody who wants to cut a deal come on down and we'll roll in the muck together a bit" (personal interview). It is essential that people engaged in any multistakeholder process assess whether the discussion is communicative action and mutual problem solving or hardball strategic negotiation. If the latter, then the participants must have the skills to operate effectively in such a forum, and organizations should fund and send representatives with skills in negotiation.

Nor was the playing field level for the handful of women who participated. The monthly two-to-three-day meetings away from home made participation difficult for women with families, women who, for the most part, are still primary caregivers. The discursive style of the proceedings may also have contributed to gender underrepresentation, but that is a topic for further investigation. One of the female participants described the style of communication at the table this way: "Sometimes some women participants were at a disadvantage because the men sitting around the table were more practised at positioning and at bluffing. There's a different way of approaching the negotiation. Men are more practised at bluffing and blustering" (personal interview). The disproportionate male presence at the table may have been because, as one participant said, "most decision makers in society are male, and that's a reality – the table reflected society in that way" (personal interview).

Debate about Substantive Issues Was Contained
A third reason for the lack of success of the Vancouver Island CORE was the constraint placed on debate about substantive issues. Most participants agreed that the deliberations had been designed by staff and facilitators to be "too process oriented" at the expense of forthright engagement in the contentious issues. The preset agenda, predetermined structure, and lack of equality among participants certainly contributed to this containment and may have preordained CORE's ultimate failure. The way that the

process managers handled conflict was also a key factor in the participants' inability to grapple with and resolve quarrelsome issues. Britell's observation that "managers or facilitators choreograph meetings so that peer-group pressure smothers substance" (1992: 21) was borne out in the Vancouver Island sessions.

One representative expressed his frustration at the lack of substantive debate this way: "These guys [the facilitators] embarked from the assumption right from the beginning that they didn't want us to engage on the major issues because we might blow apart. But at some stage that becomes totally counterproductive. We spent a gazillion meetings talking about the vision statement, a gazillion meetings talking about the participation agreement. A gazillion extremely boring sessions trading interest statements back and forth. We wasted up half a year on that. The facilitators made the assumption that, if they let us engage on the tough stuff, we would have a fight and blow up" (personal interview). Another representative reflected the same concern in a comment about the amount of time spent on issues of process: "The process could have been improved by getting the facilitator to take a more active role in forcing the pace of things and forcing the table to spend more of its time dealing with substantive issues rather than wasting so much time on insubstantive issues. That was one of my main complaints, because we spent more than 90 percent of our total table time dealing with things that were not really consequential" (personal interview).

Government-set deadlines also contributed to the table's inability to work through areas of disagreement. It is essential that participants actively engage in a discussion of differences between them if conflict resolution is to occur and that they assess and set their own realistic time line for completion of the tasks.

The Role of the Public Was Lost

A successful land-use plan depends on public support. In the case of the Vancouver Island CORE process, little support was forthcoming because the public was not involved. The interest-based sectoral structure of the multistakeholder forum precluded public involvement, and little was done by CORE staff to inform the public as the discussions proceeded. The CORE table's Policy and Procedure Agreement permitted the media to observe meetings, but there was no attempt to reach out and involve the nonaligned public through the media or any other forum. The lack of government and staff initiative in this regard was frequently mentioned. As one representative said,

The public's inclusion in this process was deficient. The public should have been invited to the process more vigorously. It was the responsibility of CORE and the government, which was after all basically funding this, to reach out to the public more energetically. The fact that, after the process was completed, many, many people in all sectors did not know what CORE's initials even stood for was astonishing to me. Each of the sectors had a responsibility to inform their sectors; wherever they could, they did. But in general there should have been eyes on the process, and there weren't. (personal interview)

Sectors were expected to communicate with their constituencies. In many cases, however, sectors were loosely defined coalitions that had never worked together before, and their extremely limited resources meant inadequate communication. Many put their efforts into involvement at the table, leaving little time and energy to communicate with others. This is another reason why the lack of adequate intervenor funding seriously compromised the outcome.

CORE lacked communications and media expertise. CORE staff had resource managers, planners, and lawyers. To its credit, CORE did not undertake a manipulative strategy of communication that could have put a falsely optimistic "spin" on the proceedings; nevertheless, a communications program could have helped to bring the public into the process. Regular contact through local newspapers and community cablevision could have provided citizens with information about the issues and the players. This approach might have motivated some citizens to enter the discussions by joining a group. Moreover, had there been some effort to encourage major media to cover CORE, there might have been wider debate and greater public acceptance of the recommendations. One representative said that in future processes there should be an effort "to raise the profile of the land-use debate away from the special interest groups and up into the public domain so it becomes sort of a main street item that people are talking about" (personal interview). Because the public was not involved in the process, there was little understanding of the complex issues and scant political will to support recommendations flowing from the deliberations. The result was that each sector took its own viewpoint to the public and to the politicians by means of staged media events after the table was dissolved. The war in the woods continued in the media.

Should Citizens Participate: Containment or True Decision Making?

Given that they can squander their limited resources, personnel, and time, citizens' groups have to think deeply before entering consensus negotiations. This study concludes that most consensus negotiations, including Vancouver Island CORE, contain debate and dissension rather than contribute to conflict resolution. As Britell has noted, "Land management agencies are making concerted efforts to undercut opposition to unpopular logging plans by getting the public more involved in planning. Agency managers have learned the value of the appearance of 'public' approval and how easy it is to get. Timber-plagued politicians are learning too, that negotiating groups are a media-friendly way to duck difficult decisions" (1992: 19).

Aware of the dangers of being contained and manipulated, potential participants must decide if conditions are present to allow some progress to be made; if those conditions are not present, they must recognize a futile containment exercise for what it is. They must further determine if process decisions are being made outside the group and if the real decision makers are participants at the table. Britell has observed that "Citizens' work groups are, almost without exception, negotiating sessions of a peculiar type: they are 'negotiations with agents with limited authority.' Negotiations of this kind present the problem that if you make concessions they are gone forever, but any gains you make can be overturned by higher authorities" (1992: 21).

In the case of CORE, decision makers were not at the table, although Commissioner Stephen Owen frequented meetings of and received reports from his staff. Since no agreement was reached at the table, Owen wrote his own report and passed on his recommendations to cabinet. The release of his report was followed by well-publicized, yellow-ribbon mass rallies and behind-the-scenes lobbying of politicians, no doubt influencing the decisions ultimately made by cabinet. Many CORE participants would echo this warning from Britell after their involvement: "Be alert for 'negotiations' which just run out the clock. By the time you realize your efforts are futile, you may have wasted the time you could have used to mobilize public opinion" (1992: 21). As a CORE participant admonished,

> Now that this CORE process is over, you should get out there and lobby like hell. You can believe that conservation will be out there lobbying, the majors are going to be leaning on the forestry doors, so we better get out there and lobby the government and the public for all of the things that we wanted in CORE. Even after all the time that we have spent in this process, we still have to do what we were doing in the first place, which is lobbying and public education. We could have saved ourselves thousands of hours and the government a couple of million dollars. (personal interview)

Such comments reflect the recognition by many participants that in the end CORE was not an essential forum for influencing key decision makers; rather, key decision makers were (and are) influenced through lobbying and media campaigns outside the process. This study and my nonrelated work experience lead me to the conclusion that media coverage and polling results are still the most compelling sources of advice for politicians.

How Citizens Can Participate

While experience and analysis of the CORE process have convinced me that citizen groups should be very sceptical about participation in multi-stakeholder forums, such processes will continue to be part of political life. Thus, it is important that myths of equality and conflict resolution be exposed so that citizens can create more inclusive shared decision-making strategies to deal with contentious issues in our society.

Multistakeholder negotiations are risky, and their results must be carefully and constantly weighed against the time and effort expended to achieve those results. However, while the conditions that promise a successful outcome rarely occur, a group may still choose to participate because it has other goals to achieve. It may decide that participation is necessary to enhance its credibility. As the youth sector representative at the Vancouver Island CORE process said, "As far as cost/benefit, I don't really think it's the most effective thing for groups to do, although I do think there needs to be people engaged in these types of activities. For youth, when there's young people getting arrested in Clayoquot, we'll be able to say we did participate in negotiating processes, and we found out what the real problems were" (personal interview).

Another reason for participating is to build a public record. Experience has shown Britell that negotiations present opportunities to build administrative records that agencies and judges cannot ignore. Some negotiations make available internal agency documentation that would otherwise be very difficult to obtain and that can become the basis for a lawsuit (1992: 23).

If citizens do decide to participate in a multistakeholder process, they can achieve the best results if they follow a disciplined course of action. They must use experienced negotiators, develop strategic goals, and make alliances to meet those goals. They must continually assess their own and other participants' strengths and weaknesses. As Britell says, "Too often, grassroots activists are emerging from advisory boards and work groups without even their slingshots intact. It doesn't have to be so. A clearer understanding of the nature of these negotiations, and careful attention to principles and details, can cut the odds environmentalists face in negotiations with land management agencies and extractive industry" (1992: 19).

Participants can increase their chances of influencing negotiations in a number of ways. They should keep the process managers to task; establish clear, task-oriented agendas; determine who are the bureaucratic and political decision makers; and insist that they attend the meetings. Participants should not allow time to be wasted on endless presentations and process items. They should ensure that official minutes are kept of meetings and that minutes be signed by participants who agree with their content and tone. They should ensure that clearly specified issues are addressed and that nonbureaucratic language is used. They should hold specific individuals accountable for undertaking tasks between meetings. Finally, they should create opportunities to have supportive scientific evidence written into the official minutes.

Overall, raising public awareness, mobilizing, grassroots organizing, lobbying government, and undertaking media work produce longer-lasting results than participating in consensus negotiations. Politicians make decisions based on electoral calculation. As one participant, who had been a municipal politician, said,

> The elected officials, particularly those on the government side of the house, and certainly cabinet, the key committees in cabinet making the hard decisions, are the ones that have to be made aware of what is going on and what the people's concerns are, and they have to be expected to respond, and if they do not I guess their tenure in government is in jeopardy. I'm not talking about three or four people writing or faxing here. I'm talking about a massive movement here. And having been involved in, albeit local, politics for some time, I know the value of hearing from people. There's nothing like a whole bunch of phone calls or letters or people screeching into your driveway to get your attention. (personal interview)

Dominant interests gain the most when a process is framed as consensus. Hence, it is often more useful for citizens' advocacy groups to continue their education and advocacy work through public forums and the media than to funnel their time and energy into multistakeholder processes such as CORE, which serve to contain rather than facilitate debate over how land should be used.

Note

1 Research for this chapter comes from Mae Burrows, "Consensus Negotiations: Conflict Resolution or Containment?" MSc thesis, School of Communications, Simon Fraser University, Burnaby, BC, 1996.

References

Amy, Douglas. 1987. *The Politics of Environmental Mediation*. New York: Columbia University Press.

Bachrach, Peter, and Morton Baratz. 1963. "Decisions and Non-Decisions: An Analytical Framework." *American Political Science Review* 57: 632-42.

Bowering, Ann (ed.). 1992. *Working Together: New Ways of Resolving Local Development Disputes*. Proceedings of a conference, 10-12 April, University of Victoria Institute for Dispute Resolution.

Brenneis, Kim, and Michael M'Gonigle. 1992. "Public Participation: Components of the Process." *Environments* 21(3): 5-11.

Britell, Jim. 1992. "Negotiate to Win." *Whole Earth Review.*

British Columbia Round Table on the Environment and the Economy. 1991. "Reaching Agreement: Volume 1 Consensus Processes in British Columbia and Case Synopses and Case Studies," Victoria.

Canada. 1994. "Evaluation and Interpretation Branch Ecosystem Conservation Directorate and Stakeholder Relations Branch Response Assessment Directorate, Environment Canada. Working in Multistakeholder Processes," Ottawa.

Commission on Resources and Environment (CORE). 1992. "Orientation Materials: Shared Decision-Making for British Columbia," Victoria.

—. 1993. "Regional Planning Guidelines: Making Regional Land Use Decisions within a Shared Decision-Making Context," Victoria.

—. 1994. "Vancouver Island Land Use Plan Volume III," Vancouver Island Table Report, Victoria, British Columbia.

Cormick, Gerald W. 1987. "The Place of Negotiation in Environmental Assessment: A Background Paper Prepared for the Canadian Environmental Assessment Research Council," Hull, PQ.

Darling, Craig. 1991. "In Search of Consensus: An Evaluation of the Clayoquot Sound Sustainable Development Task Force Process." Unpublished paper, University of Victoria Institute for Dispute Resolution.

Doering, Ronald L. 1995. "Evaluating Round Table Processes." In *The National Round Table Review.* Ottawa: National Round Table.

Flynn, Sarah Greig. 1992. "The Timber/Fish/Wildlife Agreement: A Case Study of Alternative Environmental Dispute Resolution." MSc thesis, Simon Fraser University, Burnaby, BC.

Gunton, Thomas, and Sarah Flynn. 1992. "Resolving Environmental Conflicts: The Role of Mediation and Negotiation." *Environments* 21(3): 12-21.

Lane, Patricia. 1996. "Mediation: What's New?" *Advocate* 54(2): 197-207.

National Round Table on the Environment and the Economy. 1993. *Using Consensus Processes to Promote Sustainability: Working Together for Our Common Future.* Ottawa: Canadian Round Tables.

Pearse, Peter. 1976. "Timber Rights and Forest Policy in British Columbia," Report of the Royal Commission on Forest Resources. Victoria: Royal Commission.

Pinkerton, Evelyn W. 1993. "Co-Management Efforts as Social Movements: The Tin Wis Coalition and the Drive for Forest Practices Legislation in British Columbia." *Alternatives* 19(3): 33-38.

Roseland, Mark (ed.). 1993. *From Conflict to Consensus: Shared Decision-Making in British Columbia.* Proceedings of a symposium, 5 March, Simon Fraser University Harbour Centre Campus, Vancouver.

Susskind, Lawrence, and Jeffrey Cruikshank. 1987. *Breaking the Impasse: Consensual Approaches to Resolving Public Disputes.* New York: Basic Books.

Wilson, Anne Evelyn. 1995. "Shared Decision Making in Public Land Planning: An Evaluation of the Vancouver Island Regional CORE Process." MSc thesis, Simon Fraser University, Burnaby, BC.

Wondolleck, Julia. 1988. *Conflict and Resolution: Managing National Forest Disputes.* New York: Plenum Press.

Part 5: Conclusion

10
Digging Out of the Trenches
Debra J. Salazar and Donald K. Alper

War has become a dominant metaphor used to describe forest politics in the Pacific Northwest. The war in the woods has raged over an area bounded by the fortieth and fifty-fifth parallels on the south and north and the Columbia River basin and Pacific Ocean on the east and west. There are fewer and fewer tall trees here, but they continue to define the region in important ways. Those who have made their homes here live in the land of the Douglas-fir and the western hemlock. Indeed, many of us live in houses built from the wood of these trees. We live in the rain and the fog that nurture them, and if we live in the valleys, we look out at forested slopes and up to volcanic peaks that shape the horizon.

Whether we live in cities or small towns, we are increasingly being drawn into the war in the woods. Unlike most twentieth-century wars, the battle for the coastal woods is not a clash of nation-states. Rather, it is a clash of ideas and interests in an era when the power and authority of national governments are eroding. In many ways, this war may be a prototype for twenty-first century wars. Like all wars, this one is about control of land and natural resources. But instead of national armies, the combatants are nongovernmental organizations, multinational corporations, and lone activists. The battlefields range from public hearings in government buildings, to platforms in tall trees, to global marketplaces.

Clashes in this war reflect conflicting visions of the forest. Environmentalists have challenged forest policies and practices, arguing that forests have ecological and spiritual values that outweigh their value as a source of wood products. Thus, remaining stands of ancient forests ought to be preserved, and second-growth stands ought to be managed to serve ecological rather than industrial imperatives. On the other side, representatives of wood products companies and workers have responded that severe economic disruptions will result from removing old-growth stands from the region's commercial forest base and from proscribing efficient forest practices (e.g., clear-cutting). They have contended that the massive unemployment associated with such a policy will lead to the breakdown of timber-based communities and the emergence of social pathologies such as alcoholism and domestic violence. The war in the woods has been framed as a conflict about forests, but it is more than that.

Most fundamentally, the war in the woods is about us, about how we see ourselves and our role in the world as well as how we live together and make decisions about our collective life. It reflects profound disagreements about who owns the forests and what ownership means. As well, this conflict is about democracy, about what it means for a community or a nation to govern itself in a time of global capitalism.

The perspectives taken by the contributors to this book are too different to support a grand synthesis with specific implications for forest management. Nonetheless, the authors have identified several issues at the core of forest conflict and made compelling arguments about how conflict might be resolved. Among the most important issues facing those attempting to sustain the forests of this region are the importance of monitoring forest practices for securing compliance and agreement with public goals; the structure and distribution of property rights to forest land; citizenship; and the challenges for democratic decision making in an era of global capitalism. In the remaining pages, we focus on these issues and articulate a set of criteria to guide forest policy making. These criteria derive from (1) analyses presented in the preceding chapters, (2) a presumption that forests in this region must serve a range of social purposes, and (3) our normative commitment to democratic governance.

Monitoring: Securing Compliance and Building Legitimacy

Successful implementation of forest policies requires some means of assessing and enforcing compliance. Thus, monitoring the actions of forest operators is essential to any forest program or policy initiative. The chapters by Neal Wilkins and Beverly Brown explicitly address the issue of monitoring, but the issue is implicit in several of the other chapters. There are two key questions that must be answered in designing monitoring programs. (1) What are the goals of monitoring? (2) Who will undertake monitoring? Although these questions are related, we take them up separately.

The apparent objectives of monitoring are to assess and secure compliance with rules. In implementing the Forest Practices Code in British Columbia and the forest practices acts in Oregon and Washington, public officials must have information about the extent to which forest operators are complying with the rules. Typically, compliance with regulatory rules is monitored by public officials. Seldom (if ever) is monitoring comprehensive. Public resources (and private patience) would be exhausted if comprehensiveness were attempted. Consequently, some sample of practices or outcomes is monitored and compliance assessed. Government foresters inspect harvests, roads, and tree-planting sites, then report on the number and quality of wildlife trees remaining, the adequacy of culverts, the vigour of seedlings, and other indicators of compliance. The extent of the monitoring effort and the discretion with which public foresters assess compliance vary. To the extent that forest owners and operators and activists can be included in the monitoring process, assessment can perform an educational function. Operators who participate in this process may have the opportunity to observe and reflect on the environmental consequences of various forest practices. But the key purpose of monitoring is generally seen as ensuring that operators comply with rules by creating the threat of punishment for those who fail to comply.

Monitoring also offers a means to achieve legitimacy. As Cashore and his co-authors note in Chapter 4, environmentalists are increasingly bypassing government to focus their efforts directly on forest companies. One consequence of this focus has been the emergence of forest certification programs in support of green marketing efforts. These programs use a range of criteria to certify that wood has been produced from sustainably managed forests. Of course, certification requires monitoring. Companies can then use certification as a marketing tool, targeting consumers who are willing to pay a premium to protect forest ecosystems. It is the legitimacy associated with certification that holds the promise of profiting from good management practices.[1] This legitimacy also has political value as environmental and community organizations may be less likely to challenge the forest management activities of companies that have been certified.

Certification, and its attendant market legitimacy, also expand the range of possible monitors, as certification programs may operate their own monitoring efforts. Forest industry associations and environmental organizations[2] have developed certification programs, each oriented toward a particular perspective on sustainability.[3] Industrially oriented certification programs tend to reflect an industrial model of sustained wood production, while environmentally oriented programs emphasize ecological integrity. Perhaps the next major arena of forest conflict will be a contest for public recognition of certification programs. But this contest will not be fully engaged until demonstrable financial and political gains are seen to result from forest certification.

In the interim, one of the most interesting trends in forest politics is the emergence of citizen/lay monitoring. There is a forest-monitoring network in British Columbia, and several environmental organizations in the Pacific Northwest have organized and trained citizen volunteers to monitor forest practices. Indeed, citizen monitoring may become an important mobilization tool for environmental organizations. It is in light of this trend that one should consider the argument by Neal Wilkins in Chapter 8 for self-monitoring by forest companies and the call by Beverly Brown in Chapter 6 for worker monitoring. The legitimacy of self-monitoring may be more easily defended by companies that open themselves to citizen monitoring. Similarly, workers who participate, along with citizen volunteers, in training programs jointly organized by environmental organizations and forest industry associations may also be accorded more legitimacy than if they were trained by and responsive solely to their employers. Moreover, companies may be more welcoming of citizen monitoring if the companies are allowed to participate in the training programs. These kinds of arrangements are more likely to be effective than unilateral programs.

Land Rights

Aside from differences in governmental institutions, British Columbia and the US Pacific Northwest have different property arrangements with respect to forest land. Two questions can be used to analyze property arrangements. (1) Who exercises ownership rights? (2) Which rights and duties are associated with ownership, or tenure?[4] In British Columbia, the province owns the forests, but large private companies have extensive and strong tenure claims. Thus, a key focus of forest/environmental politics has been the role of these companies and the nature of their tenures. But forestry reform proposals in British Columbia have also highlighted Native peoples and communities as alternative tenure holders. In the US Pacific Northwest, the ownership pattern is mixed. To the extent that land rights are an issue, the focus is on private land and the kinds of rights and responsibilities that private owners have with respect to the public. Contributors to this volume have addressed all of these issues. Below we review several key questions and suggest approaches to answers that are consistent with sustainability.

Perhaps the question with the greatest potential to shape forest management in British Columbia is the extent to which and the way in which Indigenous people will gain property rights to forest land. Boyd and Williams-Davidson argue in Chapter 5 that First Nations in British Columbia will practise a more ecological forestry than occurs under current tenure arrangements. There are many reasons to believe that this prediction will be realized. Boyd and Williams-Davidson describe the cultural role of forests for Indigenous peoples on the West Coast and contend that Native culture and old-growth forests must be sustained together. They identify a "traditional wisdom" about stewardship that can help to guide the transition from unsustainable industrial logging to "ecosystem-based community forestry." They note further that Canadian courts have linked Aboriginal rights claims to sustainable resource management. Thus, Boyd and Williams-Davidson make a strong case that realization of Native land rights will bring with it more ecologically based forest management and use. But there are counterpressures suggesting that Native forest management will not create an ecoforestry utopia. Unemployment on the reserves is high, and First Nations officials face internal pressures for economic development of land resources. Balancing those pressures against traditional cultural imperatives will be a major challenge for Native communities. The manner in which Native land claims are resolved will circumscribe that balancing process. Provision of external legal, financial, and technical resources may be crucial to the success of Native forestry.

Boyd and Williams-Davidson call attention to the distinction between *Aboriginal title*, an exclusive right to land, and *Aboriginal rights*, which range from resource-specific claims to title. Resolution of land claims is likely to include a mix of rights. This mix will provide an opportunity to exploit and learn from varying tenure arrangements. Native rights to fish, the subject of long legal battles in both British Columbia and the US Pacific Northwest, have crucial effects on forest management. Indeed, Native American tribes, drawing on their legal victories in fishing rights and habitat protection cases, were key to reforming forest practices in the US Pacific Northwest. Boyd and Williams-Davidson describe the analogous situation of Aboriginal rights to fish and to specific culturally significant sites in British Columbia. Such rights constrain management resources of who holds legal title to the land and create the potential for cooperative management arrangements. Cooperative management, like inclusive monitoring, creates a context for conversation among those with differing perspectives and values, a conversation that may give impetus to common purpose.[5] Thus, we would add to the Boyd and Williams-Davidson contention that Native rights to land will promote sustainable forestry a further claim – that mixed property arrangements will provide a political economic context favourable to more deliberation and thus the greater possibility of democratic experiments.

A second important question related to tenure involves the consequences of expanding tenure claims of local communities in both countries. The community forestry movement in British Columbia has generated considerable political and academic momentum. There have been experiments with increasing the local role in forest management,[6] and analysts and activists have made community tenure a cornerstone of forestry reform efforts (Burda, Gale, and M'Gonigle 1998; M'Gonigle 1998). Calls for an expanded local role in public forest management have been less common in the environmental community in the US Pacific Northwest, though, as Beverly Brown notes in Chapter 6, local governmental and nongovernmental organizations have advocated such an expansion.

The arguments for community-based forest management are compelling and draw on an extensive body of research on common property systems.[7] Local communities have stakes in perpetuating the productivity of forest resources. They also have experience with and knowledge of these resources that may not be available to bureaucrats in provincial capitals or in district offices. Thus, a greater role for community institutions in managing forests would inject local values, priorities, and knowledge into decision-making processes that too often have neglected all of these inputs. But we would echo Patricia Marchak (1998) in cautioning against tenure reform that is built solely on community control. The conditions (political, economic, biophysical) necessary for successful community management do not exist in much of the region. Many communities lack a tradition of broad citizen participation in public affairs. There is considerable inequality in the distribution of wealth and income in rural, as well as urban, areas. There are no apparent means for communities to realize income from many of the ecosystem services provided by forests. Few communities are likely to be surrounded by public forest land with a distribution of timber that would support uninterrupted harvest cycles and stable incomes. All of these factors suggest that community control is unlikely to be a universal remedy for the problems created by industrial forestry.

Moreover, as Beverly Brown observes, local control of forests will privilege those who control local political economic institutions. Brown is particularly concerned with the interests of forest-floor workers, who are often not year-round residents of a single community and whose marginal economic and social status make them unlikely to have much influence in local decision-making processes. But these are not the only people whose marginal status might be magnified in a system of local control. Thus, any effort to promote local control ought to include sensitivity to the importance of democratizing local institutions.

A third land rights issue relates to the legal claims of corporations and other private entities to forest land and timber. Will stabilizing and strengthening the tenure arrangements of corporations promote sustainability? Neal Wilkins (Chapter 8) and Clark Binkley (Chapter 7) argue that such stability is essential. Wilkins argues for a more flexible and stable regulatory structure. His account of habitat conservation efforts at Port Blakely Tree Farms suggests that such a structure will promote ecologically sensitive management. Binkley adds to this a call for stronger private rights to forest lands in British Columbia, either through long-term leases or fee simple ownership. He contends that strong and stable private property rights will promote investment in second-growth timber management and suggests that a system based on such rights is appropriate on lands designated for intensive timber management. Thus, the question becomes, how stable and how strong should private property claims be?

Several criteria are important in answering this question. First, any distribution of property rights ought to be assessed by its effect on democracy. If we are correct in asserting that sustainability is fundamentally about democracy, then property arrangements ought to be designed to support, rather than hinder, the functioning of democratic institutions. No entity's property rights should be so strong as to disenfranchise an identifiable group of citizens. Second, tenure arrangements ought to be structured to promote investment in commodity production, as well as ecosystem services, from the region's forests. The goods and services produced by the forests help to support our economies. Moreover, as Tom Waggener (Chapter 3) notes, to the extent that we prevent timber harvesting in this region, we may be displacing environmental destruction to less wealthy regions. As citizens of the world, one of our responsibilities is to find a balance between our consumption patterns and their environmental consequences. Thus, while many in the region would like to see radical changes in the structure of forest management and policy making, reformed institutions should promote investment in a regional forest economy. Third, if local and regional democracy is to be strengthened, then local and regional actors (for-profit companies as well as nonprofit community organizations) should be important in shaping forest management. One avenue of influence is tenure or ownership. If our regulatory policies make it impractical, or undesirable, for all but the largest multinational corporations to manage forests, then we might want to reconsider these policies.

Citizenship: Who Counts and How?

Sustaining the forests through democratic institutions implies an important role for citizens. Citizenship connotes a set of prerogatives and responsibilities that attaches to membership in a political community (Walzer 1988). Key issues in forest politics have related to facilitating the exercise of those rights and responsibilities and to membership in the community. We take up the issue of membership first.

Beverly Brown notes in her chapter that many forest-floor workers are neither formal citizens of the United States or Canada nor treated as citizens of the region in which they work. She argues that, if we are to expect these workers to exercise responsibility in their work in the woods, then we ought to accord them consideration as citizens. She assesses various means for achieving this, finding none of them without serious flaws. Perhaps the key initial step in securing citizenship for forest-floor workers is to acknowledge the role that they play in Pacific Northwest forest economies and ecosystems and to give them opportunities to voice their values, concerns, and interests.

The citizenship claims of Indigenous peoples are unique among those involved in forest politics. Indigenous nations are demanding sovereignty or self-government rights (Kymlicka 1996). Although Indigenous people's land claims in British Columbia are more expansive (politically and geo-graphically) than those of Indigenous people in the US Pacific Northwest, both assert a kind of right that is qualitatively different from that of others involved in forest politics. It is not just the claim of land rights but also that of political sovereignty that distinguishes the manner in which Native peoples participate in forest politics. This distinction is most apparent in the refusal of First Nations to participate in CORE as stakeholders. Rather, they insisted that any involvement on their part would be on a government-to-government basis. Kymlicka (1996) has argued that self-government rights are the most difficult kind of citizenship claim for liberal democracies to address. He argues that neither the United States nor Canada can point to success in this area. He cautions that, while recognizing self-government rights within a "multination" state may weaken the ties binding citizens to the larger political community, denying such rights will also stress community ties.

Although legal citizenship is a status generally accorded to individual persons, corporate entities often act as political citizens. Nearly every chapter in this collection, either explicitly or implicitly, examines this role. Ben Cashore and his co-authors focus specifically on forest companies and derive a set of lessons intended to promote the responsible exercise of corporate citizenship. George Hoberg notes in Chapter 2 that political decision making in British Columbia has always treated corporations as citizens, even according them privileged access to public officials. Mae Burrows argues in Chapter 9 that multistakeholder processes serve to perpetuate that privilege.

It is important to ask whether we should treat corporations as citizens. More pragmatically and immediately, we need to design decision-making processes that promote greater equality and do not privilege aggregations of capital over real persons. Providing for equal citizenship has perhaps been the most difficult challenge of liberal democracy, and there is no reason to expect the forestry arena to be more conducive than others.

The multistakeholder, collaborative process has perhaps been the most prominent innovation of natural resource policy. In an effort to promote deliberation across the battle lines, governments and nongovernmental organizations have created numerous multistakeholder processes. These processes, with varying success, have provided forums for the articulation of alternative visions of forests and forest management. Ostensibly, they are intended to promote conversation and give rise to consensus decisions. Participation has generated both cynicism and enthusiasm. Mae Burrows, drawing on her observation of the CORE process, does not convey an enthusiastic assessment.

Burrows is critical of CORE for a number of reasons. It was seen by the participants as too process oriented at the expense of substantive debate;

it was unequal in terms of participants' resources and authority; it excluded the public at large; and it was not self-governing in terms of process design or agenda setting. Multistakeholder processes cannot follow a consensus model since participants will normally have fundamentally different sets of values and goals. A strategic negotiation is possible, but consensus is not. Burrows cautions environmental and community activists to be wary of multistakeholder processes that contain debate and divert resources from grassroots organizing while significant decisions are being made elsewhere. Building public support for an issue or cause is still the most effective way to influence policy. Her conclusion suggests that equal citizenship may be promoted most effectively through civil society. That is, nongovernmental organizations and extragovernmental arenas may provide the best means to mobilize and give voice to those who have been marginalized from political conversation and public policy making.

Once again, Brown's account of the organizing efforts of forest-floor workers is suggestive. Brown notes that labour law in the United States favours employers and offers few options for contingent, temporary workers. She is generally pessimistic about the likelihood of legal changes that will empower forest-floor workers. But she sees hope in the beginnings of cross-ethnic, multilingual networking among forest workers and harvesters. Such networking may create new opportunities to challenge the structure of work in the woods and assert the citizenship claims of forest-floor workers. Moreover, an enfranchised forest-floor workforce, one that is better trained and interactive, may be better positioned to engage in "collaborative stewardship" and exercise its responsibility in the development of ecosystem management.

Global Capitalism and Democracy in the Forests

We have argued that sustainable forest use and management can occur only in a context of democratic politics. The citizen engagement and interaction characteristic of such politics are essential to forging truces in the war in the woods. In this chapter, we have drawn on the analyses of the contributors to this volume to address the challenges facing those who would pursue sustainability and democracy in the forests. Several of these contributors have noted the fundamental importance of globalization to this process. George Hoberg, Ben Cashore and his co-authors, and Tom Waggener all point to global economic and political integration as a key force in shaping the context of forest policy making and management in the region. Tom Waggener and Clark Binkley offer arguments about the importance of regional competitiveness in increasingly global markets for forests, forest products, and capital.

This competition operates as a constraint on democracy because it imposes an objective, the low-cost production necessary for competitiveness, on policy making. Public officials accountable for regional economic performance and private officials accountable for the financial performance of their companies come to see this goal as prior to all others. Multinational corporations with business interests in forest land and manufacturing pursue profit. Their control of land and capital, coupled with the dependence of regional economic performance on corporate investment and employment decisions, allows forest corporations to exercise political power in the service of profit. This power to shape or constrain decisions about the collective life of a community may be exercised directly or indirectly. Direct exercise occurs when corporations privatize decisions of public import, such as when they harvest all of the merchantable timber in a locale in a short period or modernize mills, thus severely affecting the availability of employment. Political power is exercised indirectly when companies threaten to withdraw capital, and thus jobs, from a region if disagreeable policies are enacted (Lindblom 1982).

Many in British Columbia have criticized globalization of the (forest) economy because it promotes this concentration of power in the hands of a few who are only weakly accountable to the many (Salazar and Alper 1999).[8] Lack of accountability is exacerbated by the increased international mobility of capital and the attendant decline in the capacity of national states to regulate the behaviour of multinational corporations. More specifically, critics have argued that multinational corporate control of BC forests disenfranchises the citizens of the province; no effort to transform forest management will be effective without reform of the tenure system that grants disproportionate power to these corporations (Salazar and Alper 1996). This critique has not been voiced so vigorously or so widely in the US Pacific Northwest, likely because of the structure of tenure arrangements in area forests as well as because of cultural differences in the politics of the two countries.

Regardless of how or by whom the critique is articulated, it raises an important issue. How can forest policy makers consider the range of options likely to emerge from inclusive deliberation when they are constrained by the political economic context of global capitalism? Native efforts to gain land rights, expansion of community control, inclusion of forest-floor and other workers in decision making – all seem to be doomed as too ambitious for a constrained policy space that prioritizes competitiveness in global markets over justice in local politics. Once again, our response is pragmatic. The domestic political economic role of multinational corporate entities fundamentally challenges democratic polities. Members of these polities ought to debate this role and consider the full range of alternatives for restructuring democratic politics. But as this process of deliberation unfolds, we should note that democratic experiments in forest management and policy making are proceeding.

Their scale is small, their effects are local and limited, but they do occur. Moreover, such experiments, whether the focused collaboration of diverse community organizations in northwest Washington, the organizing efforts of forest-floor workers in southern Oregon, or the operation of a log-sorting facility in interior British Columbia, promote deliberation. They facilitate conversation among diverse groups of citizens, and they inevitably highlight institutions that threaten sustainability and democracy, as well as those that support these processes.

Ending the war in the woods will require clear understanding of the sources of disagreement as well as identification of the issues for which even different perspectives yield a common vision. This kind of understanding can come out of deliberative processes only if proponents of differing views have opportunities for meaningful participation. These processes will no doubt raise issues related to monitoring, property rights, citizenship, and globalization. To the extent that these issues can be addressed by creating opportunities for all citizens to exercise their rights and responsibilities, democracy will be strengthened. Democratic decision making will be further promoted to the degree that historically marginalized groups (e.g., community-based organizations, Indigenous people, forest-floor workers) have opportunities to voice their concerns and share their knowledge effectively. The challenges are formidable. Creating monitoring systems and tenure arrangements that recognize the complexity of the forests, our multiple demands on them, and the imperatives of social justice will be no simple task. Can our political communities be sufficiently inclusive to engage all those who merit membership in them? Can we create political space for governmental and nongovernmental organizations to craft small-scale democratic experiments? These are the true tests of sustainability. We will be successful in sustaining our forests only to the extent that we can forge inclusive public conversations about our relationships with the forests and with one another.

Notes

1 This is a promise that has not necessarily been realized. See Hansen and Punches (1998).
2 In British Columbia, for example, the Silva Forest Foundation offers its services in eco-forestry management planning and certification.
3 Cashore and his co-authors (Chapter 4) discuss the relationship between the goals of certification programs and perspectives on sustainability; also see Hammond and Hammond (1997).
4 See Hallowell (1943) for a fully developed framework for comparing property arrangements. Also see Duncan (1996); Freyfogle (1996); Macpherson (1978); and Rose (1994) for varying perspectives on the relationships between private and public claims to land resources.
5 See Pinkerton's (1993) account of the Tin Wis Coalition in British Columbia.
6 See, for example, Mater and Mater's (1998) account of such an experiment in Vernon, a forest district in the interior of British Columbia. Also see Parfitt (1998) for several other examples.

7   See, for example, Bromley (1992); McCay and Acheson (1987); and Ostrom (1990).
8   There is a long tradition of such scholarship in British Columbia; examples include Marchak (1983, 1995). Analysis of political economic power and global capitalism is much less common in the US Pacific Northwest, but see Foster (1993) and, for a less academic but lively and acute critique of corporate forestry, Bari (1994).

References

Bari, Judie. 1994. *Timber Wars*. Monroe, ME: Common Courage Press.

Bromley, Daniel W. (ed.). 1992. *Making the Commons Work: Theory, Practice, and Policy*. San Francisco: Institute for Contemporary Studies.

Burda, Cheri, Fred Gale, and Michael M'Gonigle. 1998. "Eco-Forestry versus the State(us) Quo: Or Why Innovative Forestry Is Neither Contemplated Nor Permitted within the State Structure of British Columbia." *BC Studies* 119: 45-72.

Duncan, Myrl L. 1996. "Property as a Public Conversation, Not a Lockean Soliloquy: A Role for Intellectual and Legal History in Takings Analysis." *Environmental Law* 26(4): 1094-1160.

Foster, John Bellamy. 1993. "The Limits of Environmentalism without Class: Lessons from the Ancient Forest Struggle in the Pacific Northwest." *Capitalism Nature Socialism* 4(1): 11-41.

Freyfogle, Eric T. 1996. "Ethics, Community, and Private Land." *Ecology Law Quarterly* 23(4): 631-61.

Hallowell, Irving. 1943. "The Nature and Function of Property as a Social Institution." *Journal of Legal and Political Sociology* 1: 115-38.

Hammond, Herb, and Susan Hammond. 1997. "What Is Certification?" In Alan Rike Drengson and Duncan MacDonald Taylor (eds.), *Ecoforestry: The Art and Science of Sustainable Forest Use*. 196-9. Gabriola Island, BC: New Society Publishers.

Hansen, Eric, and John Punches. 1998. "Collins Pine: Lessons from a Pioneer." Case study from *The Business of Sustainable Forestry*, a project of the Sustainable Forestry Working Group. Corvallis, OR: Oregon State University.

Kymlicka, Will. 1996. "Equality, Difference, Public Representation." In Seyla Benhabib (ed.), *Democracy and Difference: Contesting the Boundaries of the Political*. Princeton: Princeton University Press.

Lindblom, Charles E. 1982. "The Market as Prison." *Journal of Politics* 44(2): 324-36.

McCay, Bonnie J., and James M. Acheson. 1987. *The Question of the Commons: The Culture and Ecology of Communal Resources*. Tucson: University of Arizona Press.

M'Gonigle, Michael. 1998. "Living Communities in a Living Forest: Ecosystem-Based Structure of Local Tenure and Management." In Chris Tollefson (ed.), *The Wealth of Forests: Markets, Regulation, and Sustainable Forestry*. 152-85. Vancouver: UBC Press.

Macpherson, C.B. (ed.). 1978. *Property: Mainstream and Critical Positions*. Toronto: University of Toronto Press.

Marchak, Patricia. 1983. *Green Gold: The Forest Industry in British Columbia*. Vancouver: UBC Press.

—. 1995. *Logging the Globe*. Montreal: McGill-Queen's University Press.

—. 1998. "Commentary." *BC Studies* 119: 73-78.

Mater, Catherine M., and Scott M. Mater. 1998. "Vernon Forestry: Log Sorting for Profit." Case study from *The Business of Sustainable Forestry*, a project of the Sustainable Forestry Working Group. Corvallis, OR: Oregon State University.

Ostrom, Elinor. 1990. *Governing the Commons: The Evolution of Institutions for Collective Action*. Cambridge, UK: Cambridge University Press.

Parfitt, Ben. 1998. *Forest Follies: Adventures and Misadventures in the Great Canadian Forest*. Madeira Park, BC: Harbour Publishing.

Pinkerton, Evelyn W. 1993. "Co-Management Efforts as Social Movements: The Tin Wis Coalition and the Drive for Forest Practices Legislation in British Columbia." *Alternatives* 19(3): 33-38.

Rose, Carol M. 1994. *Property and Persuasion: Essays on the History, Theory, and Rhetoric of Ownership*. San Francisco: Westview Press.

Salazar, Debra J., and Donald K. Alper. 1996. "Perceptions of Power and the Management of Environmental Conflict: Forest Politics in British Columbia." *Social Science Journal* 33(4): 381-99.

—. 1999. "Beyond the Politics of Left and Right: Beliefs and Values of Environmental Activists in British Columbia." *BC Studies* 121: 5-34.

Walzer, Michael. 1988. "Citizenship." In Terrence Ball, James Farr, and Russell Hanson (eds.), *Political Innovation and Conceptual Change.* 211-29. New York: Cambridge University Press.

Contributors

Donald K. Alper is Professor of Political Science and Director of the Center for Canadian-American Studies at Western Washington University. His research on BC politics and Canada-US relations is published in *BC Studies*, *American Review of Canadian Studies*, and *Canadian Public Policy – Analyse de Politiques*. He co-authored *Canada: Northern Neighbor*, which is widely used in American schools. He is active in the National Consortium for Teaching Canada, an organization that communicates the most recent research themes and ideas about Canada and its place in the world among educators at all levels.

Clark S. Binkley, Chief Investment Officer, leads the Hancock Timber Resource Group's research, client account management, and business development efforts. Immediately prior to joining HTRG, Binkley was Dean of the Faculty of Forestry at the University of British Columbia. He has written more than 100 books and articles on forest economics, and is known worldwide for his research on timberland investments.

David R. Boyd is a Senior Associate with the Eco-Research Chair in Environmental Law and Policy at the Faculty of Law, University of Victoria. He is also an Adjunct Professor at the School of Resource and Environmental Management, Simon Fraser University, and a former Executive Director of the Sierra Legal Defence Fund. Boyd is currently writing a book on environmental law in Canada, editing a forthcoming anthology of contemporary Canadian nature writing, teaching at the Hollyhock School for Environmental Leadership, and volunteering as a director for several progressive organizations.

Beverly A. Brown is the Coordinator of the Jefferson Center for Education and Research, where she is currently involved in supporting a non-timber forest contract worker and harvester networking project. She is a member of the Rural Development Leadership Network – a multicultural activist network focusing on high-poverty rural areas of North America. Brown is author of *In Timber Country: Working People's Stories of Environmental Conflict and Urban Flight*.

Mae Burrows has been the Executive Director for the T. "Buck" Suzuki Environmental Foundation and the Environmental Director for the United

Fishermen and Allied Workers' Union since 1992. She has worked with Greenpeace, the Sierra Club, the Commission on Resources and Environment (CORE), and the Rivers Defence Coalition. She currently serves as Executive Director of the Labour Environmental Alliance Society and Chair of the Environment Committee of the Vancouver District Labour Council.

Benjamin Cashore is Assistant Professor of Forest Policy and Economics at Auburn University in the School of Forestry and Wildlife Sciences. His research interests include globalization and the privatization of environmental governance in the forest sector (forest certification eco-labeling programs); comparative forest resource policy; the political economy of US/Canada forest products trade; and forest industry environmental/sustainability initiatives. Cashore is co-author of *In Search of Sustainability: The Politics of Forest Policy in British Columbia in the 1990s*, forthcoming from UBC Press.

George Hoberg is Associate Professor of Political Science and Forest Policy at the University of British Columbia. He is the author of *Pluralism by Design: Environmental Policy and the American Regulatory State*, co-author of *Risk, Science, and Politics: Regulating Toxic Substances in Canada and the United States*, co-editor of *Degrees of Freedom: Canada and the United States in a Changing Global Context*, and editor of *North American Integration: Economic, Cultural, and Political Dimensions*.

Rachana Raizada holds a PhD in International Business from the University of British Columbia. She takes a strong interest in environmental issues both at a personal and professional level. At the professional level she seeks to examine the organizational challenges to environmental policy formulation and implementation, and to uncover how these challenges can be best resolved. At the personal level she concentrates on observing and recording developments in urban transportation and related land planning issues in various countries around the world.

Debra J. Salazar is Professor of Political Science and Affiliate Professor of Environmental Studies at Western Washington University. Her research focuses on the role of social justice in the discourse and practice of environmentalism. She is particularly interested in how increasing diversity within the environmental movement is transforming environmental politics. Salazar's articles have appeared in *Social Science Quarterly*, *Society and Natural Resources*, *Natural Resources Journal*, *BC Studies*, and the *Journal of Forestry*, among others.

Ilan Vertinsky is the Director of the Forest Economics and Policy Analysis Research Unit (FEPA), Vinod Sood Professor of International Business Studies, and Professor in the Institute of Resources and the Environment at the University of British Columbia. His many published papers on forest policy deal with a variety of aspects related to questions of sustainable forest management, market access, and trade-offs between timber and other values of the forest. He is a co-editor of *Forest Policy: International Case Studies*.

Thomas R. Waggener is Professor Emeritus of Forest Economics, Policy, and International Trade at the College of Forest Resources, University of Washington. He has also served as a member of the Canadian Studies Program and the Russian, East European, and Central Asian Program of the Jackson School of International Studies at the University of Washington. He presently undertakes international consulting through International Forest Sector Analysis (IFSA) with an emphasis on forest resources development and trade in Asia and the Pacific.

R. Neal Wilkins is Assistant Professor of Wildlife and Fisheries at Texas A&M University. He previously worked as a Wildlife Biologist at Port Blakely Tree Farms, L.P., headquartered in Seattle, Washington. His position involved coordinating responses to state and federal regulations governing forest management on private lands in the Pacific Northwest. Wilkins' research focuses on functional relationships between wildlife habitat conservation and commercial forest management on private lands.

Terry-Lynn Williams-Davidson is Executive Director of the EAGLE Project (Environmental-Aboriginal Guardianship through Law and Education) of the Sierra Legal Defence Fund. She is a member of the Haida Nation from Skidegate, Haida Gwaii (Queen Charlotte Islands). She belongs to the Aboriginal, Environmental and Alternative Dispute Resolution sections of the BC branch of the Canadian Bar Association and is a board member for the Gowgaia Institute, Haida Gwaii.

Index

Note: "CORE" stands for Commission on Resources and the Environment; "ESA" for Endangered Species Act; "HCP" for habitat conservation plans; "NEPA" for National Environmental Policy Act; "NFMA" for National Forest Management Act; "SCLDF" for Sierra Club Legal Defense Fund. "(t)" after a number indicates a table, "(f)" a figure.

Aboriginal Forest Land Management Guidelines, 129
Aboriginal rights (British Columbia): *Calder* case (1973), 136, 144n50; challenge to BC forest management regime, 139-41; commercial rights to resources, 138-9, 146n94; description, 133-4, 235; entrenched in Constitution (1982), 133; fishing rights, 138, 140, 147n100, 235; forest rights, 137-8, 235; *MacMillan Bloedel* v. *Mullin* (1985), 136, 139; *Marshall* case (1999), 138, 140; rights "integral" to culture, 138; *Sparrow* case, 138, 140; *St. Catherines Milling* case (1888), 136
Aboriginal title (British Columbia): *Calder* decision (1973), 130, 136, 144n50; *Delgamuukw* decision (1997), 134-6, 140, 146n87; description, 133, 146n87; "inherent limit" on land-use activities, 134-5; interest includes forests, 136-7; proof of title, 135, 145n74. *See also* First Nations land claims; Property rights
Adams, R. v. (Canada, 1996), 139, 147n100
Alliance of Forest Workers and Harvesters (United States), 161-2
American Forests (association), 156

Annual allowable cut (AAC) (British Columbia): calculated by chief forester, 177-8; and limitations of Forest Practices Code, 126; reductions in BC and US, 47; and spotted owl case (1995-97), 46-7
Applegate Partnership, 156
Appurtenancy clauses, 185
Asia: forest product imports from Canada and US, 64-5

Babbitt, Bruce, 42
BC Treaty Commission: interim measures agreements (IMAs), 132-3; Nuu-Chah-Nulth and IMA in Clayoquot Sound, 132; shortcomings of commission, 131-2
Bentley, Peter, 92
Bilateral trade. *See* Trade in forest products
British Columbia: Commission on Resources and the Environment (*see* CORE); political culture and forest practices, 8-9. *See also* Environmentalists (British Columbia); Forest industry (British Columbia/Canada); Forest policies (British Columbia/Canada)
Bruce Babbitt, Secretary of the Interior, et al. v. *Sweet Home Chapter of Communities for a Greater Oregon et al.* (US, 1995), 15, 195
Brundtland Commission: sustainability concept (1987), 82, 112n8

Calder v. *Attorney-General of British Columbia* (1973), 130, 136, 144n50
Canada: parliamentary system, 7; responsibility for natural resources, 8. *See also* Forest industry (British Columbia/Canada); Forest policies (British Columbia/Canada)

Canadian Forest Products Limited. *See* Canfor

Canadian Reforestation and Environmental Workers Society (CREWS), 162, 165

Canadian Standards Association (CSA) Forest Program: compared with other programs, 86-7; consultation requirements, 86; forest companies, involvement, 95-6, 104; origins, 85-6

Canfor: certification sought under CSA, 95-6; description, 91; Forest Practices Performance Review Program, 95; Howe Sound pulp mill, air and water pollution, 92-3; initial response to environmental pressure (late 1980s), 92-3; proactive responses to environmental pressure, 95-6; publicity of its water pollution, 93, 94; reaction to new pollution regulations (early 1990s), 93-4; relationship with regulatory officials, 91, 92, 93; responses to pressure, compared with other companies, 107(t)

Certification programs, forest industry: Canadian Standards Association (CSA) Forest Program, 85-7, 95-6, 104; at Canfor, 95-6; Forest Stewardship Council (FSC), 84-5, 87(t); legitimatization for companies, 233; at MacMillan Bloedel, 101, 104, 115n44; Sustainable Forestry Initiative (SFI) (US), 85, 87(t), 90; at Weyerhaeuser USA, 90

CFIC (Commercial Fishing Industry Council), 210

Clark, Glen: cancellation of CORE, 33

Clayoquot Sound (British Columbia): action vs. clear-cut logging (1985), 12, 97, 114n28; cabinet decision re logging (1993), 44, 100; failure of consensus-based negotiation, 44, 45; and "free-market environmentalism," 108; Nuu-Chah-Nulth agreement with MacMillan Bloedel (1997), 103-4; protest against MacMillan Bloedel (1993-97), 100-4; scientific panel appointed (1993), 100, 102

Clear-cutting (British Columbia): a forest policy issue, 12-13; and "inherent limit" on land-use activities, 135; phased out by MacMillan Bloedel, 104-5; predominant harvesting method, 126

Clinton, Bill: forest plan (1993), 42-3; forest summit (1993), 15, 42; pro-environment administration, 15, 41-2

CMTs (culturally modified trees), 127, 143nn31-2

Coercive isomorphism, 81-2

Commercial Fishing Industry Council (CFIC), 210

Commission on Resources and the Environment. *See* CORE

Community-based forestry movement: attitude of national environmental groups, 157; collaborative groups, 156-7; contracts evolving into limited property rights, 159-60; empowerment of local businesses, 149, 154, 236; and nontimber forest workforce, 158, 160; "property right" in forest management, 162-5, 235-6; stewardship contracts and nontimber forest workforce, 159; training programs and union preferences, 160-1

CORE (Commission on Resources and the Environment) (British Columbia): agenda, process, participation predetermined, 219-20, 221, 222-3, 239; Clayoquot Sound logging protest, 100; consensus process, attempt at, 211, 212-13; consensus process, failure of, 13, 33-4, 45, 213, 221-2, 238-9; context, social and political, 209-10; creation (1992), 12-13, 210, 212; decision makers, not influenced by process, 225-6; environmental groups, non-participants, 219-20; government representation problematic, 212-13; lack of communications/media expertise, 224; lack of public participation, 223-4, 239; lack of shared values, 221-2, 239; land-use planning mandate, 14, 33, 45, 179, 212; participants, 211, 212; power imbalance of participants, 220-2, 239; substantive issues not addressed, 222-3, 239; support for environmentalists, 33

Creighton, John, 89, 90

CREWS (Canadian Reforestation and Environmental Workers Society), 162, 165

CSA Forest Program. *See* Canadian Standards Association (CSA) Forest Program

Culturally modified trees (CMTs), 127, 143nn31-2

Delgamuukw v. British Columbia (1997), 134-5, 140, 146n87

Designed Forest System (MacMillan Bloedel), 96-7

Dispute resolution processes. *See* CORE
(Commission on Resources and the
Environment); Multistakeholder
processes
Dwyer, William, 39, 41, 43

Earth Summit, Rio (UN-sponsored,
1992), 83
Ecosystem management: and community-
based forestry movement, 156-7;
different interpretations, 15-16;
expansion of NFMA, 41-2; and First
Nations culture and rights, 129,
143n43; integrative approach, 198-9;
reductionist approach, 198; sustainability of biological diversity, 82-3;
and wildlife conservation on private
lands, 198-9
Ecosystems and logging, 126-7
Endangered Species Act (ESA) (US, 1973):
definition of "harm," 195; government
enforcement, 195; habitat modification
and "incidental take," 195-7; HCPs
(*see* Habitat conservation plans, private
lands); incidental take permits, 200;
incidental take prohibition, 195; key
element in forest conflict, 14-15;
Port Blakely Tree Farms' HCP, 201-2,
203-5; on private lands (*see* Wildlife
conservation, private lands); scope
of legislation, 30; spotted owl, on
endangered list, 39. *See also* Wildlife
conservation
Environmentalists: component of policy
regime framework, 27; corporate
responses to pressures, 81-2; one side
in conflict, 4; sustainability, different
definitions, 82-3; timber harvest, effect
of environmental constraints, 75-6;
view of community-based forestry
movement, 157. *See also* Wildlife
conservation
Environmentalists (British Columbia):
actions in global marketplace, 37;
failure of litigation for old-growth
preservation, 45-6; Greenpeace
International, 37; impact of CORE on
program, 33; impact of Forest Practices
Code, 36-7; lack of mandatory legislation, 34, 38, 46; marbled murrelet
case (1991), 45-6; procedural rights,
compared with US, 36-7; Sierra Legal
Defence Fund, 34-5; spotted owl and
annual allowable cut case (1995-97),
46-7. *See also* Canfor; MacMillan
Bloedel; Wildlife conservation

Environmentalists (Pacific Northwest/
US): ESA legislation re endangered
species, 30, 39; nationalization of
old-growth debate, 40-1, 44; NEPA
and environmental impact statement,
30; NFMA and wildlife protection
regulation, 29-30; inand pluralist
legalism (policy regime), 38, 44;
Republican attempts to ease
protections (1994 onwards), 43-4;
SCLDF and wildlife protection case,
39-41; spotted owl case, effect on
global timber supply, 76; spotted
owl litigation, 39-41; victory in
Clinton's 1993 forest plan, 42-3.
See also Weyerhaeuser USA; Wildlife
conservation
ESA. *See* Endangered Species Act (ESA)
European Community: forest product
imports from Canada and US, 64-5

"Fall down effect," 177
FEMAT (Forest Ecosystem Management
Assessment Team) report, 15
First Nations and forests (British
Columbia): Aboriginal Forest Land
Management Guidelines, 129;
Aboriginal rights to forests, 133-4,
136-7, 137-9, 146n84; attempts to log
on traditional territories, 128-9; control
of, objectives, 5, 123; cultural heritage
not protected, 127; cultural importance
of forests, 123-5, 141; culturally
modified trees (CMTs), 127, 143nn31-2;
ecological impacts of logging, 126-7;
exclusion from tenure system, 127-8,
136; First Nations interests vs.
environmentalists', 12; forest
management reform, 129; Nisga'a
agreement, logging levels, 131,
145n58. *See also* First Nations land
claims (British Columbia)
First Nations land claims (British
Columbia): Aboriginal title (*see*
Aboriginal rights; Aboriginal title);
basis of logging protests, 12; BC treaty
negotiation process, 131-3; *Calder*
decision (1973), 130, 136, 144n50;
crucial importance to forest management, 121, 234-5; *Delgamuukw* decision
(1997), 134-6, 140, 146n87; fiduciary
duty of provincial government, 136;
interim measures agreements (IMAs),
132-3; litigation plus negotiation, 130;
Nisga'a Final Agreement (1998), 130,
131, 145n58; self-government rights

demanded, 237-8; unextinguished rights to most of BC, 123, 129-30, 132

First Nations Summit: call for boycott of BC timber, 129

FLs (forest licences) (British Columbia), 31

Forest Act (British Columbia), 32

Forest Appeals Commission (British Columbia), 36

Forest Development Plan (British Columbia), 35

Forest Ecosystem Management Assessment Team (FEMAT) report, 15

Forest industry: certification programs, 84-7; component of policy regime framework, 27-8; corporate case studies (*see* Canfor; MacMillan Bloedel; Weyerhaeuser USA); corporate responses, lessons learned, 105-6, 108-10; corporate responses to environmental pressure, 81-2; corporations as political citizens, 238; Forest Stewardship Council (FSC), 84-5, 87(t); "free-market environmentalism," 108-9; limited regulatory regime and company compliance, 109; monitoring for policy compliance, 232-4; one side in conflict, 4; organizational structure's importance, 110; resource depletion, stages, 175-7; "social licence" for companies, 80, 108, 112n3; standardization, binationally, 24, 54; sustainability (*see* Sustainable forestry); Sustainable Forestry Initiative (SFI) (US), 85, 87(t), 90

Forest industry (British Columbia/Canada): Aboriginal title, impact on industry, 136-7, 139-41, 146n84; Canfor (*see* Canfor); Clayoquot Sound protest (*see* Clayoquot Sound); clear-cut logging, 12-13, 104-5, 126, 135; control by few companies, 127-8; corporate responses to environmental pressures, 81-2; *Delgamuukw* decision (1997), 134-6, 140, 146n87; economic dependence on forestry, 49, 51n25; ecosystems, impact on, 126-7; ecosystems, impact on, 126-7; exports, by country, 63-5; exports, conifer logs to Japan, 69, 70(f), 71, 72(f); exports, conifer lumber to Japan, 71, 73(f), 74; "fall down effect," 177; and First Nations (*see* First Nations and forests); Free Trade Agreement (1989), 60; globalization of markets (*see* Globalization); harvest, impact of environmental constraints, 74-6; lack of account-

ability, 240; land claims, First Nations (*see* First Nations land claims); land-use allocation, 13, 14; licensing system, 11; MacMillan Bloedel (*see* MacMillan Bloedel); multiyear contracts preferred, 158-9; ownership of land, provincial, 31, 234; production compared with United States, 56, 57(t); productivity challenges, 180-1; reduced harvest from second-growth forests, 61; reduced harvests, economic and social impacts, 178, 190n5, 231-2; reforestation (*see* Reforestation); Softwood Lumber Agreement (1996), 61-2; "softwood lumber wars," 60-2; sustained yield management philosophy, 125-6; trade balances (1993-97), 66, 69(t); trade statistics, world share of Canada and US, 66, 67(t)-68(t); trade with United States, 54-5, 56, 58, 59(f), 60(f), 63-4, 65(t); unsustainable rate of logging, 126. *See also* Forest Practices Code (British Columbia); Sustainable forestry

Forest industry (Pacific Northwest/US): corporate responses to environmental pressures, 81-2; economic dependence on forestry, 49; exports, conifer logs to Japan, 69, 70(f), 71, 72(f); exports, conifer lumber to Japan, 71, 73(f), 74; exports of wood products, by country, 63-5; Free Trade Agreement (1989), 60; globalization of markets (*see* Globalization); harvest, impact of environmental constraints, 74-6; legislation governing forests, 14-15, 29-31; NFMA (*see* National Forest Management Act); ownership pattern of forests, 14-15, 28, 234; policy regime of pluralist legalism (policy regime), 28-31; political culture and forest practices, 9-11; private forests, 28; production compared with Canada, 56, 57(t); reduced harvest from second-growth forests, 61; Softwood Lumber Agreement (1996), 61-2; "softwood lumber wars," 60-2; trade balances (1993-97), 66, 69(t); trade statistics, world share, 66, 67(t)-68(t); trade with Canada, 56, 58, 59(f), 60(f), 63-4; Weyerhaeuser USA (*see* Weyerhaeuser USA)

Forest land allocation. *See* Forest policies (British Columbia/Canada); Forest policies (Pacific Northwest/US)

Forest licences (FLs) (British Columbia), 31

Forest policies: in binational region, 3, 6, 23, 26; Canada and US compared, 48-9; environmental stance and jurisdictional level, 38-9, 40; monitoring for compliance, 232-4; significance of North American forests, 55. *See also* Forest policies (British Columbia/Canada); Forest policies (Pacific Northwest/US); Globalization

Forest policies (British Columbia/ Canada): attempts at consensus-based negotiations, 13, 33-4, 44-5; Cabinet and committee domination, 44-6; Commission on Resources and the Environment (*see* CORE); compared with US, 48-9; desired outcomes for sustainability, 178; and economic dependence on forestry, 49, 51n25; executive-centred bargaining, 31-2, 44; failure of litigation for old-growth preservation, 45-6; "feedback loops" of national decisions, 77-8; Forest Practices Code (*see* Forest Practices Code); Free Trade Agreement (1989), 60; and globalization, 76-8, 239-40; lack of mandatory environmental legislation, 34, 36-7, 38, 46; NAFTA and impact of Mexico as partner, 62-3; Old-Growth Strategy Committee (1992), 44; policy reform (*see* Sustainable forestry; Tenure arrangements [British Columbia]); Protected Areas Strategy (1992), 44-5, 48; provincial jurisdiction, 8, 31; provincial ownership of land, 31, 234; Softwood Lumber Agreement (1996), 61-2; "softwood lumber wars," 60-2; stumpage charges, 31, 62; sustainability (*see* Sustainable forestry); tenure arrangements (*see* Tenure arrangements [British Columbia]); Timber Accord (1997), 62. *See also* Aboriginal rights; Aboriginal title; Clear-cutting; First Nations land claims (British Columbia); Reforestation

Forest policies (Pacific Northwest/US): Clinton administration, impact of, 15, 41-3; Clinton's forest plan (1993), 42-3; compared with Canada, 48-9; and economic dependence on forestry, 49; ecosystem management, 15-16; ESA (*see* Endangered Species Act); "feedback loops" of national decisions, 77-8; Free Trade Agreement (1989), 60; globalization, effect of, 76-8; judicialization of environmental protection, 39-44; legislation, impact of, 29-31; NAFTA and impact of Mexico as partner, 62-3; nationalization of old-growth debate by SCLDF, 40-1, 44; NEPA (*see* National Environmental Policy Act); NFMA (*see* National Forest Management Act); nondiscretionary government duties, 29; ownership pattern of forests, 14-15, 28, 234; pluralist legalism, advantage for environmentalists, 38, 44; pluralist legalism (policy regime), 28-31; private forests and public responsibilities, 16-17; Republican attempts to ease protections (1994 onwards), 43-4; Softwood Lumber Agreement (1996), 61-2; "softwood lumber wars," 60-2. *See also* Environmentalists (Pacific Northwest/US)

Forest policy regime frameworks: categories of actors, 27-8; causal and normative beliefs, 27-8; institutions (rules and procedures), 27-8. *See also* Forest policies (British Columbia/ Canada); Forest policies (Pacific Northwest/US)

Forest Practices Board (British Columbia), 36

Forest Practices Code (British Columbia): codification of regulations, 13, 35; Forest Appeals Commission, 36; Forest Development Plan (FDP), 35; Forest Practices Board, 36; framework for planning, 35-6; in land-use zoning scenario, 181-2; limitations to power, 126; operational plans, 35-6; Silviculture Prescription (SP), 35

Forest Renewal BC (FRBC), 183-4, 191n9

Forest Sector Strategy (British Columbia), 13, 14

Forest Stewardship Council (FSC) (British Columbia): compared with other programs, 84-5, 87(t)

Forests: conflicting visions of, 4, 231-2; North American, significance in world, 55; resource depletion, stages, 175-7

Franklin, Jerry, 100

FRBC (Forest Renewal BC), 183-4, 191n9

Free Trade Agreement (1989): and softwood lumber disputes, 60

"Free-market environmentalism," 108-9

FSC (Forest Stewardship Council) (British Columbia), 84-5, 87(t)

Gladstone, R. v. (1996), 138, 147n100

Globalization (market integration): and forest policies (BC), 76-8, 239-40;

global demand for timber (2010, 2020), 75; pressures, 83; timber and international competition, 62, 65-6; timber supply/demand and environmental constraints, 75-6; trade statistics, world share, 66, 67(t)-68(t). *See also* Internationalization; Trade in forest products

Governments: Canadian and American systems, 7-8, 24, 26, 27; component of policy regime framework, 27-8; multistakeholder processes (*see* CORE; Multistakeholder processes); political culture and forest practices, 8-11

Greenpeace: actions in global marketplace, 37; agreement with MacMillan Bloedel, 104-5, 115n46; campaign for chlorine-free pulp, 98, 99; Clayoquot Sound campaign, 100-1, 102

Greenpeace Germany campaigns, 94, 101

Habitat conservation plans, private lands: authorized under ESA, 88, 113n17, 200-1; landowners' willingness to participate, 194, 199, 205-6; monitoring process, 203-5; multispecies plans, 172, 200; of Port Blakely Tree Farms' plan, 201-2, 203-5; of Weyerhaeuser USA, 90

Haida First Nation: action on South Moresby archipelago (1985), 12, 97; lack of local benefits from logging, 127

Harcourt, Mike, 12-13, 101, 212

Heiltsuk First Nation, 138

Howe Sound Pulp and Paper Limited. *See* Canfor

IMAs (interim measures agreements), 132-3, 145n61

Incidental take permits (under ESA), 200

Incidental take prohibition (of ESA), 195

Indigenous peoples. *See* Aboriginal rights; Aboriginal title; First Nations and forests (British Columbia); First Nations land claims (British Columbia)

Interim measures agreements (IMAs), 132-3, 145n61

Internationalization: Earth Summit (UN-sponsored, 1992), 83; effect on national forest policies, 23-4, 54-5, 76-8, 239-40; "feedback loops" of national decisions, 77-8. *See also* Globalization

Isomorphism (acquiescence to outside pressure), 81-2

Japan: conifer log imports, 69, 70(f), 71, 72(f); conifer lumber imports, 71, 73(f),

74; importance in forest products trade, 66, 69; wood imports from Canada and US, 64-5

Kerr, Andy, 40

Kitlope (British Columbia): protected area, 45

Land claims, native or First Nations. *See* First Nations land claims (British Columbia)

Land and Resource Management Plans (LRMPs) (British Columbia), 34, 45

Land Use Coordination Office (British Columbia), 34

Land-use zoning (British Columbia): careful implementation required, 188-9; defined by CORE, 179; and ecosystem functions of forests, 171-2; and Forest Practices Code, 181-2, 188, 189; intensive timber production areas, 179-80, 181, 182; joint-use areas, 181; to maintain harvest levels, 179-80; to protect environmental quality, 181-2; protected areas, 181

Lead Partnership Group (LPG), 156

MacMillan Bloedel: agreement with Greenpeace, 104-5, 115n46; Carmanah Valley logging protest (1988), 98-9; certification efforts, 101, 104, 115n44; Clayoquot Sound protest (1993-97), responses, 100-4; clear-cutting to be phased out, 104-5; company description, 96-7; development of Designed Forest System, 96-7; initial response to environmental pressure (1970s-1980s), 97-9; Meares Island in Clayoquot Sound, logging attempt (mid-1980s), 97; NDP election in 1991, effect of, 99, 102; new CEO appointed, 104-5; Nuu-Chah-Nulth, logging agreement for Clayoquot Sound (1997), 103-4; proactive responses to environmental pressures (1998 and later onwards), 104-5; purchased by Weyerhaeuser USA, 110-11; responses to pressure, compared with other forest companies, 107(t); South Moresby Island logging protest (1985), 12, 97, 114n28; sustainability, company concept, 97; vice-president of environmental affairs, 101-2, 115n45

MacMillan Bloedel v. *Mullin* (Canada, 1985), 136, 139

Marshall v. *R.* (Canada, 1999), 138

Meares Island (Clayoquot Sound): logging attempt by MacMillan Bloedel, 97, 114n28

Mexico: lumber issues and NAFTA, 62-3

Mi'kmaq First Nation, 138

Mimetic isomorphism, 82

Ministry of Forests Act (British Columbia), 32

Monitoring forest practices: certification programs, 233; collaborative groups and stewardship, 157-8; legitimatization for companies, 233; objectives, 232-3; self- and citizen-monitoring, 233-4. *See also* Certification programs, forest industry; Habitat conservation plans, private lands

Multistakeholder processes: agenda to be developed by participants, 214-15; citizen groups, decision whether to participate, 226, 239; citizen groups, influencing negotiations, 226-7, 239; different values and goals to be recognized, 218-19; differing views of, 238-9; evaluation of Commission on Resources and the Environment (*see* CORE); funding to be provided, 216-17; mainly containment exercises, 224-6; participation to be as equals, (inclusive or restricted scenario), 215-16; power to be balanced, 217-18; process to be participant-designed, 214

NAFTA (North American Free Trade Agreement) (1994): and lumber policies, 62-3

National Aboriginal Forestry Association: Aboriginal Forest Land Management Guidelines, 129

National Environmental Policy Act (NEPA) (US, 1969): environmental impact statement requirement, 30; listing spotted owl as endangered, 39; purpose, 14

National Forest Management Act (NFMA) (US, 1976): case of endangered spotted owl, 39-41; planning and regulation, 29-30; purpose, 14; wildlife protection regulation, 29-30

National forests (United States): responsibility of US Forest Service, 28-9. *See also* Old-growth forests (Pacific Northwest/US)

Native land claims. *See* Aboriginal rights; Aboriginal title; First Nations land claims (British Columbia)

Native peoples. *See* Aboriginal rights; Aboriginal title; First Nations and forests (British Columbia); First Nations land claims (British Columbia)

NEPA. *See* National Environmental Policy Act (NEPA) (US)

New Democratic Party (British Columbia): actions to reduce forest policy conflict, 12-13, 32; loosened government-forest industry ties, 93-4, 99, 102, 114n30; revised environmental regulations (1991), 94, 99

New Zealand, 187, 188

NFMA. *See* National Forest Management Act (NFMA) (US)

Nisga'a Final Agreement (1998), 130, 131, 145n58

Nontimber forest workforce: Canadian Reforestation and Environmental Workers Society (CREWS), 162, 165; and community-based forestry (*see* Community-based forestry movement); contract labour and harvesting, similarities, 152; discrimination, 161, 164, 165; diversity, 148-9, 150-1; enfranchisement, 237; lack of employer accountability, 152-3, 159; lack of rights, status, or visibility, 5, 149, 152-3, 154-5; products increasing in importance, 122, 156; "property right" to participate in forest management, 163-5; under stewardship contracts, 158-9; training programs and union preferences, 160-1; US Alliance of Forest Workers and Harvesters, 161-2; work arrangements, types, 151-2, 153-5

Normative isomorphism, 82

North American Free Trade Agreement (NAFTA) (1994): and lumber policies, 62-3

Northern spotted owl: and annual allowable cut case in BC, 46-7; effect of case on global timber supply, 76-7; environmental case in US, 39, 41-2; habitat conservation plans (1992 onwards), 200; Port Blakely Tree Farms' HCP, 201-2, 203-5

Nuu-Chah-Nulth First Nation: joint logging venture in Clayoquot Sound, 103-4, 132; protest over Carmanah Valley logging, 99; stewardship ethic toward land, 124-5

Old-growth forests (British Columbia): amount remaining, 48; clear-cutting phased out by MacMillan Bloedel, 104-5; conflicting visions of, 231-2; failure of environmental litigation, 45-6; and First Nations culture, 123-5; percentage protected, 47, 48; restrictions on protections, 48

Old-growth forests (Pacific Northwest/US): amount remaining, 48; forest plan (Clinton, 1993), 42-3, 47; percentage protected, 42, 47; Republican attempts to ease protections, 43-4; spotted owl case, 39, 41-2, 47

Old-Growth Strategy Committee (British Columbia, 1992), 44

Oregon, 6, 14-15. *See also* Forest policies (Pacific Northwest/US)

Oregon Natural Resources Council, 40

Owen, Stephen, 33, 212, 225

Pacific Northwest (United States). *See* Forest policies (Pacific Northwest/US); Oregon; Washington

Pluralist legalism: forest policies in Pacific Northwest/US, 28-31

Political culture and forest practices: in British Columbia, 8-9; in Pacific Northwest/US, 9-11

Port Blakely Tree Farms: habitat conservation plan, 201-2, 203-5

Property rights: Aboriginal (*see* Aboriginal rights; Aboriginal title; First Nations land claims); community forestry's "property rights," 162-5, 235-6; evolution from multiyear contracts, 159-60; private lands and wildlife conservation, 196; stronger rights proposed with tenure reform, 184-5, 189, 191n10, 236-7. *See also* Tenure arrangements (British Columbia)

Protected Areas Strategy (1992) (British Columbia), 44-5, 48, 181

Pulp mill pollution: dioxins and shellfish industry (1988), 93, 98; regulations re effluent (1992), 94, 99; regulations re organochlorines (1989), 93

Reforestation: backlog in BC in early 1980s, 12; Designed Forest System of MacMillan Bloedel, 96-7; silviculture, 35, 179-80, 188; sustained yield management, 125-6; Weyerhaeuser USA as pioneer, 88

Russia: conifer log exports to Japan, 69, 70(f), 71, 72(f); conifer lumber exports to Japan, 71, 73(f), 74; response to US supply cutbacks, 76, 78n3

Scientific Panel for Sustainable Forest Practices in Clayoquot Sound, 100, 102

SFI (Sustainable Forestry Initiative, American Forest and Paper Association), 85, 87(t), 90

Shellfish industry closures, 93, 98

Sierra Club Legal Defense Fund (SCLDF) (United States), 39-41

Sierra Legal Defence Fund (British Columbia), 34-5, 50n6

Silviculture, 179-80, 188. *See also* Reforestation

Silviculture Prescription (SP) of Forest Practices Code, 35

Smith, Ray, 98

"Social licence" for forest companies, 80, 108, 112n3

Softwood lumber: "lumber wars" (1960s onwards), 60-2; Softwood Lumber Agreement (Canada and US, 1996), 61-2

South Moresby archipelago: Haida action vs. clear-cut logging (1985), 12, 97

Sparrow case, 138

Spotted owl. *See* Northern spotted owl

St. Catherines Milling and Lumber Co. v. *The Queen* (Canada, 1888), 136

Stephens, Tom, 104-5

Stewardship approach to land: habitat conservation plans (*see* Habitat conservation plans, private lands); land-use zoning (*see* Land-use zoning [British Columbia]); multistakeholder process (*see* Multistakeholder processes); and nontimber forest workforce, 158-9; Nuu-Chah-Nulth First Nation belief, 124-5; by Weyerhaeuser USA, 89-91

Sto:lo First Nation, 138

Stumpage charges (British Columbia), 31, 62

Sustainable forestry: compensation for losses due to policy changes, 185-6, 189; concept of, and response to environmental pressures, 109-10, 111; different definitions, 82-3; First Nations approach to forests, 141; introduced by Brundtland Commission (1987), 82, 112n8; key element, enhancing productivity, 180-1; key element, maintaining environmental quality, 181-2; key element, maintaining harvest levels, 178-80, 182; land-use zoning (*see* Land-use zoning); MacMillan Bloedel's concept of, 97; policy adjustments required, 182-3, 188; political concept, 6; research and development needed, 186-8; technology-based strategy essential, 186-8; tenure reform required, 183-5; unsustainable rate of logging in BC, 126; Weyerhaeuser USA's belief in, 87-8

Sustainable Forestry Initiative (SFI), American Forest and Paper Association, 85, 87(t), 90

Sustained yield: forest management philosophy (BC), 125-6

Suzuki (T. "Buck") Environmental Foundation, 210

Sweet Home Chapter of Communities for a Greater Oregon et al., Bruce Babbitt, Secretary of the Interior et al. v. (US, 1995), 15, 195

T. "Buck" Suzuki Environmental Foundation, 210

Tasman Forest Accord (New Zealand), 188

Tenure arrangements (British Columbia): absent from CORE's agenda, 221; appurtenancy clauses, 185; description, 31, 183, 185; First Nations excluded, 127-8; forest licences (FLs), 31; Forest Renewal BC (FRBC) and investment capital, 183-4, 191n9; reform not addressed by Forest Sector Strategy, 14; reform with stronger property rights, 184-5, 189, 191n10, 236-7; timber supply areas (TSAs), 31; tree farm licences (TFLs), 31

TFLs (tree farm licences) (British Columbia), 31

Thomas, Jack Ward, 42

Timber supply areas (TSAs) (British Columbia), 31

Timber-dependent rural communities: coalitions in Pacific Northwest/US, 5-6

Trade in forest products: with Asia, 64-5; Canada-US, 56, 58, 59(f), 60(f), 63-4, 65(t); Canada-US-Mexico (under NAFTA), 62-3, 64(t); with European Community, 64-5; "feedback loops" of national decisions, 77-8; Free Trade Agreement (1989), 60; global demand for timber (2010, 2020), 75; international statistics, Canada and US compared, 66, 67(t)-68(t); Japanese log imports, by country, 69, 70(f), 71, 72(f); Japanese lumber imports, general pattern, 71, 73(f), 74; Japan's importance, 66, 69; North America largest market, 54; Softwood Lumber Agreement (1996), 61-2; "softwood

lumber wars," 60-2; timber supply, impact of environmental constraints, 75-6; trade balances, Canada and US (1993-97), 66, 69(t)

Tree farm licences (TFLs) (British Columbia), 31

Tree planters. *See* Nontimber forest workforce

TSAs (timber supply areas) (British Columbia), 31

Union of BC Indian Chiefs, 129

United States: forest industry, conflicting policy interests, 4-5; government system, 7-8; responsibility for natural resources, 8, 28; trade with Canada in forest products, 54-5. *See also* Forest industry (Pacific Northwest/US); Forest policies (Pacific Northwest/US)

US Forest Service: ecosystem protection, ruling by Dwyer, 41-3; national forests (*see* National forests [United States]); NFMA, and spotted owl protection case, and (SCLDF), 39-41; NFMA and wildlife protection, 29-30; "Thomas report," 41

USSR. *See* Russia

Van der Peet, R. v. (Canada, 1996), 138

Washington (state), 14-15. *See also* Forest policies (Pacific Northwest/US)

Watershed councils (Oregon), 6

Weyerhaeuser USA: belief in sustainability and high-yield forestry, 87-8; globalization, 91; habitat conservation plans (mid- to late 1990s), 90; initial response to environmental pressures (1960s-1980s), 88-9; intensive silviculture, 179; new CEO appointed (1989), 89, 90; responses to pressure, compared with other forest companies, 107(t); stewardship approach ("Weyerhaeuser Forestry"), 89-91; support of Sustainable Forestry Initiative (SFI), 90; takeover of MacMillan Bloedel, 110-11

Wildlife conservation, federal lands: spotted owl case (BC, 1995-97), 46-7; spotted owl case (US, 1987-94), 39-41. *See also* Endangered Species Act (ESA); National Environmental Policy Act (NEPA); National Forest Management Act (NFMA)

Wildlife conservation, private lands: absolute habitat thresholds, 196; competing interests of landowners,

193-4; definition of "harm," 195; ecosystem management approach, 198-9; habitat modification and incidental take, 195-7; HCPs (habitat conservation plans), 88, 113n17, 199-201, 236-7; HCPs, monitoring process, 203-5; HCPs, Port Blakely Tree Farms, 201-2, 203-5; incidental take permits, 200; incidental take prohibition, 195; legal controversy re "taking" of property, 194; protection of individual animals, not populations, 196-7; unlisted species, protection disincentives, 197-8